Christian Hollauer

Modeling of Thermal Oxidation and Stress Effects

AF135251

Christian Hollauer

Modeling of Thermal Oxidation and Stress Effects

with the Finite Element Method

Südwestdeutscher Verlag für Hochschulschriften

Impressum/Imprint (nur für Deutschland/ only for Germany)
Bibliografische Information der Deutschen Nationalbibliothek: Die Deutsche Nationalbibliothek verzeichnet diese Publikation in der Deutschen Nationalbibliografie; detaillierte bibliografische Daten sind im Internet über http://dnb.d-nb.de abrufbar.
Alle in diesem Buch genannten Marken und Produktnamen unterliegen warenzeichen-, marken- oder patentrechtlichem Schutz bzw. sind Warenzeichen oder eingetragene Warenzeichen der jeweiligen Inhaber. Die Wiedergabe von Marken, Produktnamen, Gebrauchsnamen, Handelsnamen, Warenbezeichnungen u.s.w. in diesem Werk berechtigt auch ohne besondere Kennzeichnung nicht zu der Annahme, dass solche Namen im Sinne der Warenzeichen- und Markenschutzgesetzgebung als frei zu betrachten wären und daher von jedermann benutzt werden dürften.

Verlag: Südwestdeutscher Verlag für Hochschulschriften Aktiengesellschaft & Co. KG
Dudweiler Landstr. 99, 66123 Saarbrücken, Deutschland
Telefon +49 681 37 20 271-1, Telefax +49 681 37 20 271-0, Email: info@svh-verlag.de
Zugl.: Vienna, Technical University, PhD, 2007

Herstellung in Deutschland:
Schaltungsdienst Lange o.H.G., Berlin
Books on Demand GmbH, Norderstedt
Reha GmbH, Saarbrücken
Amazon Distribution GmbH, Leipzig
ISBN: 978-3-8381-0532-1

Imprint (only for USA, GB)
Bibliographic information published by the Deutsche Nationalbibliothek: The Deutsche Nationalbibliothek lists this publication in the Deutsche Nationalbibliografie; detailed bibliographic data are available in the Internet at http://dnb.d-nb.de.
Any brand names and product names mentioned in this book are subject to trademark, brand or patent protection and are trademarks or registered trademarks of their respective holders. The use of brand names, product names, common names, trade names, product descriptions etc. even without a particular marking in this works is in no way to be construed to mean that such names may be regarded as unrestricted in respect of trademark and brand protection legislation and could thus be used by anyone.

Publisher:
Südwestdeutscher Verlag für Hochschulschriften Aktiengesellschaft & Co. KG
Dudweiler Landstr. 99, 66123 Saarbrücken, Germany
Phone +49 681 37 20 271-1, Fax +49 681 37 20 271-0, Email: info@svh-verlag.de

Copyright © 2009 by the author and Südwestdeutscher Verlag für Hochschulschriften Aktiengesellschaft & Co. KG and licensors
All rights reserved. Saarbrücken 2009

Printed in the U.S.A.
Printed in the U.K. by (see last page)
ISBN: 978-3-8381-0532-1

Abstract

THERMAL OXIDATION is one of the most important process steps in semiconductur fabrication, which is used to produce high quality insolation layers. The chemical reaction during oxidation converts silicon into silicon dioxide. The formed oxide material has more than twice of the original volume of silicon. This significant volume increase is the main source for stress and displacements in the oxidized structure. There is a big interest in the simulation of oxidation, because the volume increase and the fact that the oxide growth rate depends on a number of parameters and also on the stress in the material, make it impossible to predict the final shape of the silicon dioxide without simulation in practically used structures. Furthermore, the possible stress distribution and deformation which are caused by the oxidation process in the neighboring structure, can be only evaluated by simulation.

All conventional models are based on the moving boundary concept. Unfortunately moving boundaries are the most restricting factor for three-dimensional oxidation simulation, because they need complicated algorithms and an enormous data update. Therefore, a modern three-dimensional oxidation model should be based on a new concept which avoids the difficulties and drawbacks regarding the mechanics. An up-to-date model should also enable the simulation of even complex structures within an acceptable time period on convential computers. Furthermore, for universal application an oxidation model should be physically based, which means that it takes into account that thermal oxidation is a process where a diffusion, a chemical reaction, and a volume increase occur simultaneously. In the course of this work an advanced three-dimensional oxidation model which is able to fulfill all listed requirements, was developed. This model is based on a diffuse interface concept.

The implementation of the model in a simulation tool is an important task. The numerical solving of the mathematical formulation is performed with the finite element method which is most suitable for the mechanical displacement problem. The discretization of

ABSTRACT

the (differential) equations is an important part of modeling. For the practical application of the simulation tool a simple method for the model calibration is shown. Despite the diffuse interface concept, the simulation results can be presented with a sharp interface between silicon and silicon dioxide for a physical interpretation. It is known that stress has a significant influence on the oxidation growth rate. For obtaining physically meaningful simulation results, the stress dependence of the oxidation process is taken into account in the oxidation model.

Stress in copper interconnects is an important promoting factor for electromigration. The material transport due to electromigration can lead to void formation in the interconnect. These voids can cause an enormous increase of the resistance or even a total failure in the interconnect. Thermal stress arises from the self-heating effect of the current flow in the interconnect, because the copper interconnects are embedded in materials with different thermal expansion coefficients. A stress simulation is the only possible way to determine high-stress areas in the interconnect structure in order to locate critical points with respect to electromigration.

During the fabrication of micro-electro-mechanical systems and aftermath, where thin film deposition is a widely used technique, an intrinsic stress is generated in the layers. In micro-electro-mechanical systems which are mostly used as sensors, the stress can change the electrical and magnetic characteristics and can also cause unwanted deformation in free standing structures. The determination of intrinsic stress in thin films is demanded, but a number of microscopic effects which lead to stress do not allow a straightforward stress calculation. In this work a number of intrinsic stress sources are discussed. For the different intrinsic stress sources, which describe the stress development due to the microscopic effects, macroscopic mechanical formulations are given. Furthermore, a methodology which allows to predict the stress distribution in the deposited thin film, was developed.

Acknowledgment

FIRST AND FOREMOST I want to thank my superviser Prof. SIEGFRIED SELBERHERR that he gave me the chance to join his research group and to take a doctoral program at the Institute for Microelectronics. I am very grateful that he guided and supported me and my scientific work through all the years. Further, I am thankful that Prof. SELBERHERR gave me the necessary time and freedom to finish all my scientific activities successfully.

I am also indebted to Prof. ERASMUS LANGER, the head of the Institute for Microelectronics. Prof. LANGER has always been a very cooperative head and he has done a very good adiminstration. He let me take part in international research projects which was an important experience for me.

My thanks also go to Dr. HAJDIN CERIC and Dr. ANDREAS HÖSSINGER who bootstrapped me at the Institute. With HAJDIN I had a lot of valuable discussions during all the years and he has been a pool of constructive ideas and elaborated suggestions. Furthermore, he proofread this thesis and gave good comments.

Since joining the Instiute Dr. ALIREZA SHEIKHOLESLAMI has been my room mate. He has been a very pleasant and very helpful colleague. With him I had a lot of interesting discussions, lots of funny moments, and so many lunches at Mensa. He also proofread this thesis.

I would also like to thank STEFAN HOLZER for his cooperative nature, especially in case of the electro-thermal simulations.

In case of several computer system or network problems Dr. JOHANN CERVENKA was always willing and able to help. Furthermore, here I would like to say thank you to all my colleagues throughout my time at the Institute, because all of them were always very friendly and helpful. There was always a very pleasant atmosphere.

Finally I thank my parents and my girlfriend Elisabeth for their never ending support during my time at the Institute.

Contents

Abstract   i

Acknowledgment   iii

Contents   v

1 Introduction   1
- 1.1 Semiconductor Fabrication Processes   2
  - 1.1.1 Lithography   2
  - 1.1.2 Etching   2
  - 1.1.3 Deposition   3
  - 1.1.4 Chemical Mechanical Planarization   3
  - 1.1.5 Oxidation   3
  - 1.1.6 Ion Implantation   3
  - 1.1.7 Diffusion   4
- 1.2 Isolation Techniques   4
  - 1.2.1 Local Oxidation of Silicon   4
  - 1.2.2 Shallow Trench Isolation   5
- 1.3 Overview and History of Process Simulators for Oxidation   6
- 1.4 Outline of the Thesis   8

CONTENTS

2 Physics of Thermal Oxidation   11

- 2.1 The Material Silicon Dioxide 12
  - 2.1.1 Properties of SiO_2 12
  - 2.1.2 Structure of SiO_2 13
- 2.2 Principles of the Oxidation Process 14
- 2.3 Rapid Thermal Oxidation 16
- 2.4 Oxidation Parameters 17
  - 2.4.1 Oxidant Species 17
    - 2.4.1.1 Dry Oxidation 18
    - 2.4.1.2 Wet Oxidation 18
    - 2.4.1.3 Mixed Flows of O_2 with H_2O, HCL, and Cl_2 18
  - 2.4.2 Influence of Temperature 21
  - 2.4.3 Influence of Pressure 23
  - 2.4.4 Influence of Crystal Orientation 23
- 2.5 Nitrided Oxide Films 25
  - 2.5.1 Different Nitridation Methods 25
  - 2.5.2 Diffusion-Barrier Properties of Nitrided Layers 26
  - 2.5.3 Nitrogen Incorporation by NO 26
  - 2.5.4 Nitrogen Incorporation and Removal by NO_2 27
  - 2.5.5 Nitridation in N_2 and NH_3 28
- 2.6 The Deal-Grove Model 29
  - 2.6.1 Concept and Formulation 29
  - 2.6.2 Analytical Oxidation Relationship 31
  - 2.6.3 Temperature Dependence of B and B/A 33
  - 2.6.4 Pressure Dependence of B and B/A 34
  - 2.6.5 Dependence of B and B/A on Crystal Orientation 35
  - 2.6.6 Thin Film Oxidation with Deal-Grove Model 36
- 2.7 The Massoud Model 36
  - 2.7.1 Experimental Fitting 37
  - 2.7.2 Analytical Oxidation Relationship 38

3 Advanced Oxidation Model — 41
- 3.1 The Diffuse Interface Concept 42
- 3.2 Mathematical Formulation 42
  - 3.2.1 Oxidant Diffusion 42
  - 3.2.2 Dynamics of η 43
  - 3.2.3 Volume Expansion of the New Oxide 44
  - 3.2.4 Diffusion Coefficient and Reaction Layer 44
  - 3.2.5 Mechanics 46
    - 3.2.5.1 Elastic Mechanical Model 46
    - 3.2.5.2 Visco-Elastic Mechanical Model 48
    - 3.2.5.3 Volume Increase and Mechanics 50
- 3.3 Model Overview 50

4 Oxidation of Doped Silicon — 53
- 4.1 Dopant Redistribution 54
- 4.2 Five-Stream Dunham Diffusion Model 55
  - 4.2.1 Interaction of Dopants 55
  - 4.2.2 Continuity Equations 56
- 4.3 Segregation Interface Condition 57
- 4.4 Model Overview with Coupled Dopant Diffusion 58

5 Discretization with the Finite Element Method — 59
- 5.1 Basics 59
  - 5.1.1 Mesh Aspects 60
  - 5.1.2 Shape Function 60
  - 5.1.3 Weighted Residual Method 61
- 5.2 Discretization with Tetrahedrons 62
  - 5.2.1 Shape Functions for a Tetrahedron 62
  - 5.2.2 Coordinate Transformation 65
  - 5.2.3 Differentiation in the Normalized Coordinate System 67
  - 5.2.4 Discretization of the Oxidant Diffusion 69

CONTENTS

    5.2.5   Discretization of the η-Dynamics 74

    5.2.6   Discretization of the Mechanics 75

  5.3   Assembling and Solving 79

    5.3.1   Principle of Assembling 79

    5.3.2   Dirichlet Boundary Conditions 80

    5.3.3   Mechanical Interfaces 81

    5.3.4   Complete Equation System for Oxidation 83

    5.3.5   Solving with the Newton Method 84

6 Simulation of Thermal Oxidation with FEDOS   87

  6.1   Architecture of FEDOS 88

    6.1.1   Inputdeck 88

    6.1.2   Wafer-State-Server 90

    6.1.3   QQQ-solver 90

  6.2   Simulation Procedure 91

  6.3   Meshing Aspects 93

  6.4   Sharp Interface and Smoothing 96

    6.4.1   Segment Splitting 96

    6.4.2   Smoothing 98

  6.5   Model Calibration 100

    6.5.1   Calibration and Parameter Extraction 100

    6.5.2   Calibration Concept and Example 101

  6.6   Comparison with a Two-Dimensional Simulation 104

7 Stress Dependent Oxidation   105

  7.1   Oxidation Modeling with Stress 105

  7.2   Stress Calculation Concept for Simulation 107

  7.3   Representative Examples 110

    7.3.1   First Example 111

    7.3.2   Stress Dependence 111

    7.3.3   Second Example 115

CONTENTS

8 Thermo-Mechanical Stress in Interconnect Layouts — 119
- 8.1 Simulation Procedure 120
  - 8.1.1 Electro-Thermal Simulation 121
  - 8.1.2 Thermo-Mechanical Stress Simulation 122
- 8.2 Demonstrative Example 123
  - 8.2.1 Simulation Results 125
    - 8.2.1.1 Temperature Distribution 125
    - 8.2.1.2 Pressure Distribution 125

9 Intrinsic Stress Effects in Deposited Thin Films — 131
- 9.1 Cantilever Deflection Problem 132
  - 9.1.1 Principle of Cantilever Deflection 132
  - 9.1.2 Stress Distribution and Relaxation 134
- 9.2 Sources of Intrinsic Stress 135
- 9.3 Modeling of the Stress Sources 138
- 9.4 Investigation of Fabricated Cantilevers 140
  - 9.4.1 Cross Section 140
  - 9.4.2 Strain Curve 141
  - 9.4.3 Practical Example 143

10 Summary and Conclusions — 145

Bibliography — 149

Chapter 1

Introduction

THE SIMULATION of semiconductor processes is used to reduce the development time and costs of new semiconductor products, because it can replace a number of time-consuming and expensive experiments. Usually time is a critical factor in the semiconductor industry, because, if a company can bring a new product earlier to the market, it has a big advantage in competition and can make more profit.

The strength of simulation tools is that after modeling and calibration the effects of changing process parameters, materials, and geometries can be predicted in a fast and simple way. The key for accurate simulation results and all-purpose simulation tools are physically based models. All important process steps, as listed in Section 1.1, which influence the topology and characteristics of a device significantly are worth for modeling and simulation. One of these process steps is thermal oxidation.

Modeling of thermal oxidation has a long tradition. Already in the middle of the 60's the Deal-Grove model has been developed which is still used in modern oxidation simulators. The model is based on two parameters, the so-called linear and parabolic rate constant, in which all the physics of the oxidation process is included. The rate constants must be determined by experiments for the respective oxidant species. Later in the 80's, the Deal-Grove concept has been extended with additional fitting parameters, in order to describe thin oxide films.

The modeling of stress sources and simulation of its effects in semiconductor devices and micro-electro-mechanical systems becomes more and more important. Stress in a material or structure can lead to various negative or undesirable effects. During the fabrication process it can influence the physics of a process in a unpredictable way. Stress can also impair the electrical characteristics of a device and even reduce the life-time of an integrated circuit. In micro-electro-mechanical systems, which are mainly used as sensors, stress can not only change the electrical and magnetic characteristics, it can also cause unwanted deformation in a free standing structure.

INTRODUCTION

The continuously shrinking device dimensions in the state of the art ultra large scale integration (ULSI) technology brings up three-dimensional effects which can not be investigated with two-dimensional simulations. However, the industry is still often confined to use two-dimensional process simulation tools, because of missing three-dimensional alternatives. Therefore, the development of universal three-dimensional models is the actual challenge in process simulation.

1.1 Semiconductor Fabrication Processes

Starting with an uniformly doped silicon wafer, the fabrication of integrated circuits (IC's) needs hundreds of sequential process steps. The most important process steps used in the semiconductor fabrication are [1]:

1.1.1 Lithography

Lithography is used to transfer a pattern from a photomask to the surface of the wafer. For example the gate area of a MOS transistor is defined by a specific pattern. The pattern information is recorded on a layer of photoresist which is applied on the top of the wafer. The photoresist changes its physical properties when exposed to light (often ultraviolet) or another source of illumination (e.g. X-ray). The photoresist is either developed by (wet or dry) etching or by conversion to volatile compounds through the exposure itself. The pattern defined by the mask is either removed or remained after development, depending if the type of resist is positive or negative. For example the developed photoresist can act as an etching mask for the underlying layers.

1.1.2 Etching

Etching is used to remove material selectively in order to create patterns. The pattern is defined by the etching mask, because the parts of the material, which should remain, are protected by the mask. The unmasked material can be removed either by wet (chemical) or dry (physical) etching. Wet etching is strongly isotropic which limits its application and the etching time can be controlled difficultly. Because of the so-called under-etch effect, wet etching is not suited to transfer patterns with sub-micron feature size. However, wet etching has a high selectivity (the etch rate strongly depends on the material) and it does not damage the material. On the other side dry etching is highly anisotropic but less selective. But it is more capable for transfering small structures.

1.1.3 Deposition

A multitude of layers of different materials have to be deposited during the IC fabrication process. The two most important deposition methods are the physical vapor deposition (PVD) and the chemical vapor deposition (CVD). During PVD accelerated gas ions sputter particles from a sputter target in a low pressure plasma chamber. The principle of CVD is a chemical reaction of a gas mixture on the substrate surface at high temperatures. The need of high temperatures is the most restricting factor for applying CVD. This problem can be avoided with plasma enhanced chemical vapor deposition (PECVD), where the chemical reaction is enhanced with radio frequencies instead of high temperatures. An important aspect for this technique is the uniformity of the deposited material, especially the layer thickness. CVD has a better uniformity than PVD.

1.1.4 Chemical Mechanical Planarization

Processes like etching, deposition, or oxidation, which modify the topography of the wafer surface lead to a non-planar surface. Chemical mechanical planarization (CMP) is used to plane the wafer surface with the help of a chemical slurry. First, a planar surface is necessary for lithography due to a correct pattern transfer. Furthermore, CMP enables indirect pattering, because the material removal always starts on the highest areas of the wafer surface. This means that at defined lower lying regions like a trench the material can be left. Together with the deposition of non-planar layers, CMP is an effective method to build up IC structures.

1.1.5 Oxidation

Oxidation is a process which converts silicon on the wafer into silicon dioxide. The chemical reaction of silicon and oxygen already starts at room temperature but stops after a very thin native oxide film. For an effective oxidation rate the wafer must be settled to a furnace with oxygen or water vapor at elevated temperatures. Silicon dioxide layers are used as high-quality insulators or masks for ion implantation. The ability of silicon to form high quality silicon dioxide is an important reason, why silicon is still the dominating material in IC fabrication.

1.1.6 Ion Implantation

Ion implantation is the dominant technique to introduce dopant impurities into crystalline silicon. This is performed with an electric field which accelerates the ionized atoms or molecules so that these particles penetrate into the target material until they come to rest because of interactions with the silicon atoms. Ion implantation is able to control

INTRODUCTION

exactly the distribution and dose of the dopants in silicon, because the penetration depth depends on the kinetic energy of the ions which is proportional to the electric field. The dopant dose can be controlled by varying the ion source. Unfortunately, after ion implantation the crystal structure is damaged which implies worse electrical properties. Another problem is that the implanted dopants are electrically inactive, because they are situated on interstitial sites. Therefore after ion implantation a thermal process step is necessary which repairs the crystal damage and activates the dopants.

1.1.7 Diffusion

Diffusion is the movement of impurity atoms in a semiconductor material at high temperatures. The driving force of diffusion is the concentration gradient. There is a wide range of diffusivities for the various dopant species, which depend on how easy the respective dopant impurity can move through the material. Diffusion is applied to anneal the crystal defects after ion implantation or to introduce dopant atoms into silicon from a chemical vapor source. In the last case the diffusion time and temperature determine the depth of dopant penetration. Diffusion is used to form the source, drain, and channel regions in a MOS transistor. But diffusion can also be an unwanted parasitic effect, because it takes place during all high temperature process steps.

1.2 Isolation Techniques

Thermal grown oxide is mainly used as isolation material in semiconductor fabrication. For the isolation of neighboring MOS transistors there exist two techniques, namely Local Oxidation of Silicon and Shallow Trench Isolation. The differences in their process flow and their final oxide shapes are described in the following.

1.2.1 Local Oxidation of Silicon

Local **O**xidation of **S**ilicon (LOCOS) is the traditional isolation technique. At first a very thin silicon oxide layer is grown on the wafer, the so-called pad oxide. Then a layer of silicon nitride is deposited which is used as an oxide barrier. The pattern transfer is performed by photolithography. After lithography the pattern is etched into the nitride. The result is the nitride mask as shown in Fig. 1.1a, which defines the active areas for the oxidation process. The next step is the main part of the LOCOS process, the growth of the thermal oxide. After the oxidation process is finished, the last step is the removal of the nitride layer. The main drawback of this technique is the so-called bird's beak effect and the surface area which is lost to this encroachment. The advantages of LOCOS fabrication

1.2 Isolation Techniques

are the simple process flow and the high oxide quality, because the whole LOCOS structure is thermally grown.

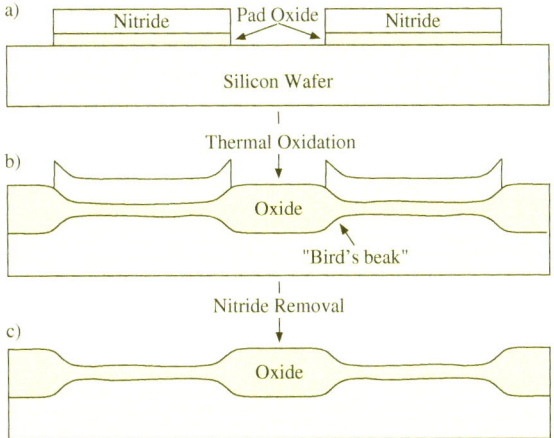

Figure 1.1: Process sequence for local oxidation of silicon (LOCOS).

1.2.2 Shallow Trench Isolation

The Shallow Trench Isolation (STI) is the preferred isolation technique for the sub-0.5 μm technology, because it completely avoids the bird's beak shape characteristic. With its zero oxide field encroachment STI is more suitable for the increased density requirements, because it allows to form smaller isolation regions. The STI process starts in the same way as the LOCOS process. The first difference compared to LOCOS is that a shallow trench is etched into the silicon substrate, as shown in Fig. 1.2a. After underetching of the oxide pad, also a thermal oxide in the trench is grown, the so-called liner oxide (see Fig. 1.2c). But unlike with LOCOS, the thermal oxidation process is stopped after the formation of a thin oxide layer, and the rest of the trench is filled with a deposited oxide (see Fig. 1.2d). Next, the excessive (deposited) oxide is removed with chemical mechanical planarization. At last the nitride mask is also removed. The price for saving space with STI is the larger number of different process steps.

INTRODUCTION

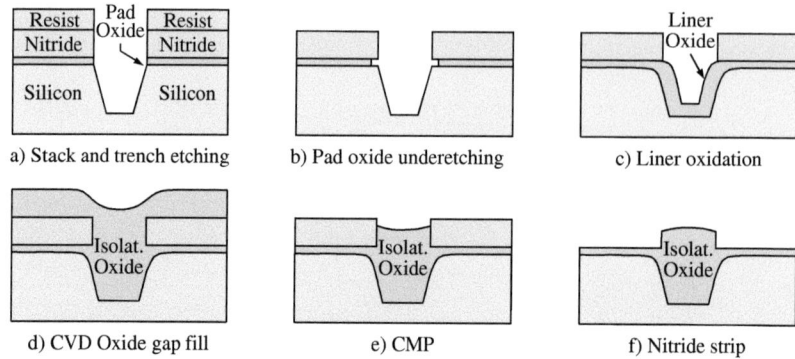

Figure 1.2: Steps in a typical shallow trench isolation (STI) process flow.

1.3 Overview and History of Process Simulators for Oxidation

Simulation of oxidation has a long tradition and a lot of people and institutions were active in oxidation modeling, also the Institute for Microelectronics [2]. Because over the decades a large number of oxidation simulators has been developed, this section is only focused on commercial process simulation tools. The history of the commercial tools is close with the history of the respective TCAD company. In principle all the following listed tools use the Deal-Grove concept (see Section 2.6) with its two rate constants and moving boundaries. The main reason to use the Deal-Grove model is the existence of the calibrated rate constants for a multitude of different oxidation conditions, because since Deal and Grove 1965 a lot of other oxidation experiments has been done. The price for this convenience are the difficulties in handling moving boundaries, especially in three dimensional geometries.

SUPREM-IV is a two-dimensional process simulator [3], which was developed at the Stanford University (Department of Electrical Engineering) in the TCAD group of Prof. Robert Dutton [4], with SUPREM a pioneer in TCAD. For oxidation SUPREM-IV has a compress and a viscous mechanical model [5]. The compress model treats the oxide as compressible liquid, while the viscous model treats the oxide as an incompressible viscous liquid. SUPREM-IV, the successor of the one-dimensional version SUPREM-III, is the basis for the two commercial tools TSUPREM-IV and ATHENA.

TSUPREM-IV [6] was the commercial version of SUPREM-IV from the company Technology Modeling Associates Inc. (TMA). TMA was founded out of Stanford University 1979 with Prof. Dutton as director and started the commercial TCAD business [7]. Approximately

1.3 Overview and History of Process Simulators for Oxidation

20 years later, in 1998 TMA was acquired by the 1986 founded company Avant! Corp. In the last Avant! release 2002 TSUPREM-IV offers for oxidation a compress, viscous, and also a visco-elastic model [6]. Last release, because in 2002 Avant! itself was acquired by the company Synopsys Inc.

ATHENA [8] is the commercial version of SUPREM-IV from the private company Silvaco International [9], which still distributes ATHENA. Silvaco was founded 1984 by Dr. Ivan Pesic [10] and is since this time successful on the TCAD market. ATHENA was never extended to three dimensions and thus it is still only a two-dimensional tool [11]. ATHENA also has the same compress and viscous mechanical models like the university version SUPREM-IV.

In 1992 TMA started a project for a new three-dimensional process simulator which is mainly based on the level-set algorithm. After TMA was acquired, Avant! released this product 1998 with the name TAURUS [12]. The mechanics during oxidation is described with a visco-elastic model. Because of the problems with moving boundaries in three dimensions, TAURUS has never become a complete stable three-dimensional process simulator [13].

The Integrated Systems Laboratories at the ETH Zurich also developed a two-dimensional process simulator named DIOS [14], which came out 1992. Later, in December 1993 the company Integrated Systems Engineering AG (ISE) was founded as a spin-off of the university laboratories. Since this time ISE distributed DIOS as a commercial tool. In the last 2004 ISE TCAD release, a viscous, elastic, or visco-elastic model for the mechanical problem can be applied. This was the last release, because in 2004 ISE was also acquired by Synopsys.

In 1993 a first version of the Florida Object-Oriented Process Simulator (FLOOPS) was completed. FLOOPS was developed at the University of Florida in the Electrical Engineering Department of Prof. Mark Law [15]. Already 1996 the PhD student Stephen Cea presented that FLOOPS has been extended from two to three dimensions and three-dimensional oxidation simulation can be performed [16]. Since 1996 the work has been continued to reach a stable three-dimensional tool, because even the actual FLOOPS version is still a little buggy for three-dimensional oxidation [17]. The 2002 release of FLOOPS was commercialized by ISE in the same year and henceforth promoted as the next generation three-dimensional process tool [18]. With the additional developments of ISE the so-called FLOOPS-ISE became a stable three-dimensional oxidation simulator. FLOOPS-ISE has the same mechanical models as DIOS (viscous, elastic, and viscoelastic), but extended for three-dimensional structures [19].

The company Synopsys Inc. [20] was founded in 1986. After Synopsys acquired Avant! and ISE, it holds now the licenses for all former Avant! and ISE tools and has with 80% market share nearly a monopoly in the TCAD market [13]. Silvaco is the only remaining competitor. For two-dimensional process simulation Synopsys sells now the packages DIOS

INTRODUCTION

and TAURUS-TSUPREM-IV [21]. After merging with ISE, Synopsys started to transfer the best features of DIOS, TAURUS and TSUPREM-IV to the FLOOPS-ISE platform for generating a new three-dimensional process simulator [22]. The first release of the new simulator with the name SENTAURUS [23] was carried out in 2005 [24].

1.4 Outline of the Thesis

The topic of this thesis are the three-dimensional modeling and simulation of thermal oxidation, which is the first and main part, and the three-dimensional modeling and simulation of stress and its effects in the second part.

In the beginning of *Chapter 2* the characteristics, properties, and structure of the material silicon dioxide and the principle of the oxidation process are described. Next the influence of the different oxidation parameters on the oxidation process are lighted up. Some aspects of nitrided oxide films are also listed. At last the concept of the traditional oxidation modeling, which is still used in state of the art oxidation simulators with more or less extensions, is explained.

In *Chapter 3* an advanced oxidation model with an effective and improved modeling concept is presented. The new concept avoids the drawbacks of the traditional oxidation models, especially regarding the mechanics in case of complex three-dimensional structures. This chapter includes mainly the mathematical formulation of the advanced oxidation model, where the mechanics is an essential part. Thermal oxidation of doped silicon material leads to a redistribution of the dopands as described in *Chapter 4*.

Chapter 5 treats the discretization of the mathematical formulation with the finite element method which starts with some basics. This chapter concentrates on the discretization with tetrahedrons, which is explained at first in general and then in detail for the used differential equations of the advanced oxidation model and the mechanics. The chapter continues with the description of the assembling procedure, also for the needed special cases like mechanical interfaces, in order to built-up the complete equation system. At the end the solving of this equation system with the Newton method is described.

Chapter 6 is focused on the simulation of thermal oxidation with the in-house process simulation tool into which the models were implemented. The architecture and main components of this tool are depicted and the simulation procedure for oxidation is explained. Since not only the accuracy, but also the simulation time and computer resources depend on the number of discrete elements, the used mesh plays a key role for simulation. So in this chapter an effective meshing strategy is discussed. Furthermore, the procedure for the sharp interface interpretation of the displayed simulation results is described. Finally the optimal way found for the model calibration is shown.

1.4 Outline of the Thesis

In *Chapter 7* the developed oxidation model is applied for stress dependent oxidation. A universal stress calculation concept for the oxidation simulation is presented. In order to demonstrate the good performance of the model and the simulation tool, representative examples for oxidation are presented.

Because stress is a promoting factor for electromigration, in *Chapter 8* the simulation procedure of thermo-mechanical stress in copper interconnect structures is described. The stress distribution for a demonstrative interconnect layout is simulated.

In *Chapter 9* intrinsic stress effects in deposited thin films are discussed. At the beginning a typical effect, the cantilever deflection problem, is shown. Furthermore, some stress sources are listed and a macroscopic stress formulation is given. A strain curve predicted by the methodology is analyzed and calibrated for a multilayer film. The calibrated curve is applied to investigate a fabricated cantilever structure. The thesis is concluded with a summary in *Chapter 10*.

Chapter 2

Physics of Thermal Oxidation

THERMAL OXIDATION is a chemical process, where silicon dioxide (SiO_2) is grown in an ambient with elevated temperatures. A simple form of thermal oxidation even takes place at room temperature, if silicon is exposed to an oxygen or air ambient. There, a thin native oxide layer with 0.5–1 nm will form on the surface rapidly. After that, the growth slows down and effectively stops after a few hours with a final thickness in the order of 1–2 nm, because the oxygen atoms have too small energy at room temperature to diffuse through the already formed oxide layer.

SiO_2 is used to isolate one device from another, to act as gate oxide in MOS structures, and to serve as a structured mask against implant of dopant atoms. In the beginning of this chapter is described, why thermal grown SiO_2 is the most suitable material for such requirements.

This chapter will focus on thermal oxidation, but it should be mentioned that SiO_2 layers can also be produced by deposition techniques, like chemical vapor deposition. Deposition normally involves a much smaller thermal budget than thermal oxidation and so it is the only option when wafers have already metal on them. Usually deposited oxides are not used for thin layers under 10 nm because the control of the deposition process is not so good as the thermal oxidation process. Another disadvantage is the interface between a deposited oxide and the underlying silicon, which is electrically not so good as thermal oxide. Furthermore, deposited oxide does not have the same high density as thermal grown oxide.

Thermal oxidation is a complex process where a diffusion of oxidants, a chemical reaction, and a volume increase occur simultaneously to convert the silicon substrate into SiO_2. This process is strongly influenced by the used oxidant species, the oxidation ambient with temperature and pressure, and also the crystal orientation of the substrate. With these parameters the quality and the growth of the oxide during the manufacturing process can be controlled.

The small dimensions and high performance of modern MOS devices require ultrathin SiO_2 layers for gate dielectrics. Apart from the exact thickness control, pure SiO_2 has some difficulties to fulfill all requirements at such thin thicknesses. Especially the dopant penetration and direct tunneling for ultrathin oxides can not be handled. It was found that silicon oxynitrides are more suitable materials for such applications. Oxynitrides can be produced by different methods which depend on the desired nitrogen profile and, therefore, on the application.

2.1 The Material Silicon Dioxide

SiO_2 is one of the most important and attractive materials in semiconductor fabrication, especially for MOS technology. In contrast to other materials which suffer from one or more problems, SiO_2 offers a lot of desired characteristics and advantages [25, 26]:

- SiO_2 layers are easily grown thermally on silicon or deposited on many substrates.
- They are resistant to most of the chemicals used in silicon processing and yet can be easily patterned and selectively etched with specific chemicals or dry etched with plasmas.
- They block the diffusion of dopants and many other unwanted impurities.
- The interface that forms between silicon and SiO_2 has very few mechanical or electrical defects and is stable over time.
- SiO_2 has a high-temperature stability (up to 1600 °C) indispensable for process and device integration.
- SiO_2 is an excellent insulator with a high dielectric strength and wide band gap.

2.1.1 Properties of SiO_2

In Table 2.1 some important properties of SiO_2 are listed [27]. The density of thermally grown dry oxide is a little bit higher than of wet oxide, which leads to a better oxide quality. The thermal expansion coefficient is a measure of stress or strain, which the oxide exerts on other materials in contact with it, practically during high-temperature cycles. The Young's modulus and Poisson's ratio describe the mechanical behavior of oxide films. In contrast to silicon the stiffness of SiO_2 is approximately only a third.

The thermal conductivity is an important parameter which affects power during circuit operation. The stability of H_2O under high electric fields is expressed as its dielectric strength which is related to the high resistivity. The bandgap of SiO_2 is nearly 8 times wider compared to the bandgap of silicon. The wide bandgap and the high dielectric strength make oxide films very suitable for dielectric isolation.

Table 2.1: Important properties of SiO_2.

Density (thermal, dry/wet)	$2.27/2.18$ g/cm^{-3}
Thermal expansion coefficient	$5.6 \cdot 10^{-7}$ 1/K
Young's modulus	$6.6 \cdot 10^{10}$ N/m^2
Poisson's ratio	0.17
Thermal conductivity	$3.2 \cdot 10^{-3}$ W/(cm·K)
Relative dielectric constant	3.7 – 3.9
Dielectric strength	10^7 V/cm
Energy bandgap	8.9 eV
DC resistivity	$\approx 10^{17}$ Ω·cm

2.1.2 Structure of SiO_2

SiO_2 can be described as a three-dimensional network constructed from tetrahedral cells, with four oxygen atoms surrounding a silicon atom [25], as shown in a two-dimensional projection in Fig. 2.1a. The silicon atoms are in the center of each of the tetrahedra. The length of a Si-O bond is 0.162 nm and the normal distance between oxygen ions is 0.262 nm. The Si-Si bond distance depends on the particular form of SiO_2 with about 0.31 nm. The six-membered ring structure of SiO_2 is shown in Fig. 2.1b. In an ideal network the vertices of the tetrahedra are joined by a common oxygen atom called a bridging oxygen.

In the amorphous forms of SiO_2 there can be also some non-bridging oxygen atoms. These phases are often named as fused silica. Crystalline forms of SiO_2 such as quartz contain only binding oxygen bonds. The various crystalline and amorphous forms of SiO_2 arise because of the ability of the bridging oxygen bonds to rotate, allowing the position of one tetrahedron to move with respect to its neighbors. This same rotation allows the material to lose long-range order and hence become amorphous. The rotation and the capability to vary the angle of the Si-SiO_2-Si bond from 120° to 180° with only a little change in energy play an important role in matching amorphous SiO_2 with crystalline silicon without breaking bonds [28].

PHYSICS OF THERMAL OXIDATION

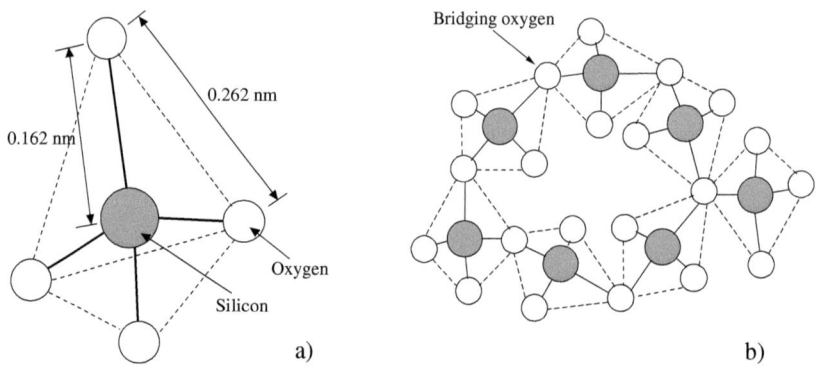

Figure 2.1: Structure of fused silica glass a) and structure of SiO$_2$ b).

2.2 Principles of the Oxidation Process

Thermal oxidation is a process where silicon is converted into SiO$_2$ with the help of oxidants in an artificial high-temperature ambient. As oxidant source different oxygen compounds can be used which are supplied by the ambient. Fig. 2.2 illustrates that the oxidants diffuse from the oxide surface through the already existing oxide to the interface. At the interface the chemical reaction takes place where the silicon is converted into the SiO$_2$ [29].

In the interface the silicon is converted in principle atom layer after atom layer. In the interface there is a mixture of silicon, oxygen, and SiO$_2$. The interface thickness is only a few atom layers. Because of the silicon consumption the interface moves constantly from the surface into the silicon substrate during the oxidation process. On the other side the molecule density of SiO$_2$ is with 2.3×10^{22} molecules/cm^3 less than half of the atom density of silicon.

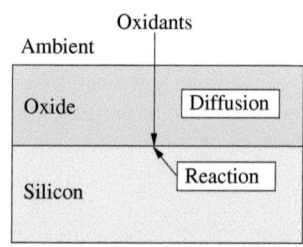

Figure 2.2: Basic process for the oxidation of silicon.

2.2 Principles of the Oxidation Process

Because of the different molecule densities of silicon and SiO_2, the newly formed SiO_2 has 125% volume expansion. If the volume expansion takes place only in one direction, as shown in Fig. 2.3, the thickness of the SiO_2 is 225% compared to the original silicon. Without mechanical boundary conditions the SiO_2 would like to expand by 31% in all three dimensions to accomodate the oxygen atoms.

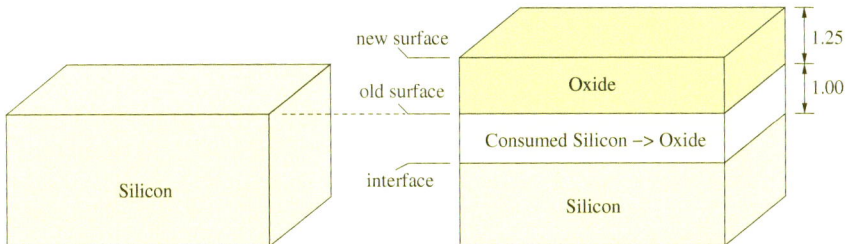

Figure 2.3: Moving interface and volume expansion.

In practice not the whole silicon surface is oxidized. So the areas which should not be oxidized are masked, usually by a silicon nitride mask, because this mask prevents the oxidant diffusion to the underlying silicon layer. The oxidants can not diffuse through silicon nitride, because compared with silicon or oxide this material has a high density. However, the oxidation process does not stop at the edge of the mask, because the oxidants are able to diffuse through the already existing oxide into regions under the mask and react there with silicon (see Fig. 2.4). The finally oxide regions are therefore normally larger than the not masked ones, but there are also natural mechanisms which strongly restrict or nearly stop the oxidation process under the mask. In the end the form of the oxide is close to the shape of the mask.

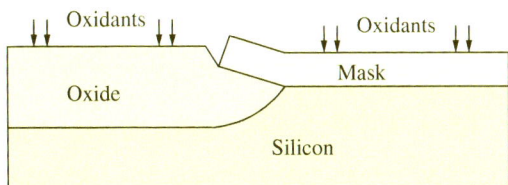

Figure 2.4: Defining the oxidation area by masking.

PHYSICS OF THERMAL OXIDATION

2.3 Rapid Thermal Oxidation

The decreasing size of the semiconductor devices demands very short high-temperature oxidation steps, because thermal oxidation influences the distribution of impurities in the bulk of silicon and at the Si/SiO_2 interface. Since the movement of impurities affects the device size and its electrical properties, it is important to control and minimize the effects of oxidation on the impurity profile. This can be achieved by precisely controlling the oxidation temperature and reducing the thermal budget of the heat cycle required for an oxide film growth.

Unfortunately, for such applications there is a limitation of the conventional furnace oxidation due to its inertia to temperature transitions, which results in a higher thermal budget than required for oxidation. The thermal budget can be reduced considerably by decreasing the duration of these transitions. As shown in Fig. 2.5, a smaller thermal budget can be achieved by rapid thermal processing (RTP) [30].

During RTP, the wafer is rapidly heated from a low to a high processing temperature (T > 900 °C). It is held at this elevated temperature for a short time and then brought back rapidly to a low temperature. Typical temperature transition rates range from 10 to 350 °C/s, compared with about 0.1 °C/s for furnace processing. So RTP reduces the ramp-up and ramp-down durations. The RTP durations at high processing temperatures vary from 1 s to 5 min. This makes RTP very suitable to grow thin oxide films (< 40 nm), where a precise temperature control and short oxidation times are important.

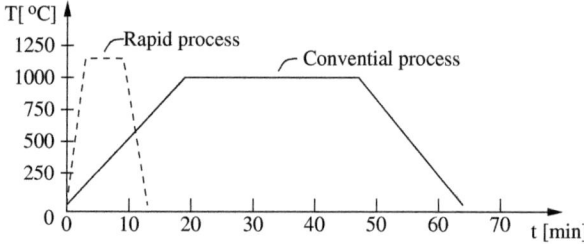

Figure 2.5: Reduction of the thermal budget with rapid thermal processing.

A schematic RTP system is shown in Fig. 2.6. The heat source is typically an array of lamps in an optical system. In contrast to conventional furnaces, where a batch of wafers is introduced into the furnace and oxidized at the same time, RTP systems are single-wafer machines, and only one wafer is in the chamber and processed. However, due to the high processing temperature (T > 900 °C), the processing time required for oxidation is in RTP systems reduced.

One of the difficult problems in an RTP system is to know exactly the wafer temperature. These systems usually support the wafer on a small thermal mass in order to heat the wafer rapidly. This makes it very difficult to use thermocouplers for temperature measurement as is done in a furnace. Another technique is to measure with an infrared pyrometer from the back side of the wafer. Precise temperature measurement is rather difficult with this method, because the "energy reading" depends mainly on the surface condition of the back side. Furthermore, the wafer temperature can change by approximately 1000 °C in a few seconds, which also complicates an accurate temperature measurement.

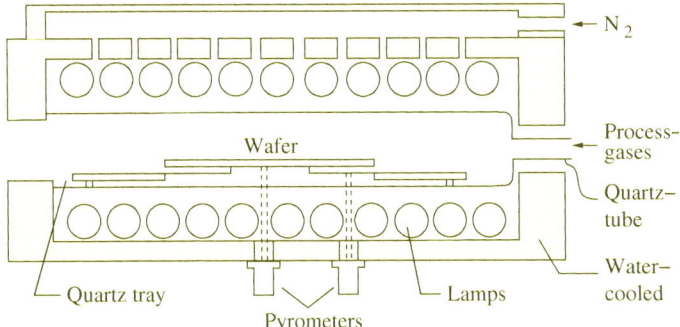

Figure 2.6: Cross section view of a RTP system.

2.4 Oxidation Parameters

The desired characteristics and requirements of the fabricated oxide can be mainly influenced by the used oxidant species. For a chosen oxidant species the oxide growth rate usually is controlled by the temperature. Additionally, it is possible to vary the hydrostatic pressure in the reaction chamber, if the oxidation system offers such possibilities. Furthermore, the oxidation rate is also influenced by the crystal orientation of the used silicon substrate.

2.4.1 Oxidant Species

The most important characteristic of oxidant molecules is that they contain oxygen atoms, which are needed for the transformation from silicon to SiO_2. The classical oxidant species are pure oxygen, which is also declared as dry oxidation, and water vapour, which is also declared as wet oxidation. In the middle of the 70's people started to mix pure oxygen mostly with Chlorine or Hydrocloric Acid to improve oxide quality and speed up growth rate. The state of the art are nitrided oxides for MOS-gates, which are in principle also produced by dry oxidation. Because of their extension and importance this species is described separately in Section 2.5

PHYSICS OF THERMAL OXIDATION

2.4.1.1 Dry Oxidation

During dry oxidation the silicon wafer is settled to a pure oxygen gas atmosphere (O_2). The oxidation rate is low (< 100 nm/hr) and so the final oxide thickness can be controlled accurately. Compared with other oxides the dry oxide has the best material characteristics and quality. The chemical reaction between silicon (solid) and oxygen (gas) is

$$Si + O_2 \rightarrow SiO_2. \tag{2.1}$$

With dry oxidation normally high quality thin oxide films up to 100 nm thickness are produced. Dry oxides are especially used as gate oxides in MOS technology. The actually fabricated gate oxide thickness is in the magnitude of about only 2 nm in the currently used 90 nm process technology, whereas the exact thickness depends on the respective manufacturing setup. Unfortunately, at such thicknesses SiO_2 generated from pure oxygen does not fulfill all demands for a good gate oxide.

2.4.1.2 Wet Oxidation

During wet oxidation the silicon wafer is settled to a water vapour atmosphere (H_2O). Wet oxides grow really fast compared to dry oxidation, which is the biggest advantage. The reason for the much higher growth rate is the oxidant solubility limit in SiO_2, which is much higher for wet (H_2O) than for dry oxidation (O_2). For 1000 °C the typical solubility limit value is 5.2×10^{16} cm^{-3} for dry oxidation compared to 3×10^{19} cm^{-3} for wet oxidation, which is nearly 600 times higher.

Therefore, wet oxidation is applied for thick oxides in insulation and passivation layers, where thick oxide buffers are needed to suppress electric currents or to ensure high threshold voltage of parasitic transistors. The chemical reaction is

$$Si + 2 H_2O \rightarrow SiO_2 + 2 H_2. \tag{2.2}$$

Because of its water content, wet oxide films exhibit a lower dielectric strength and more porosity to impurity penetration than dry oxides. Therefore, wet oxidation is used when the electrical and chemical properties of the film are not critical.

2.4.1.3 Mixed Flows of O_2 with H_2O, HCL, and Cl_2

The gas flow of O_2 can be mixed in the furnace with H_2O, HCL, and Cl_2 to get acceptable oxide quality at a higher growth rate. Besides a higher growth rate, Hydrocloric Acid (HCL) or Chlorine (Cl_2) is often used in oxidation in order to prevent metallic contamination and to help avoiding defects in the oxidation layer [31]. HCL and Cl_2 have a cleaning effect of the furnace as well as an improvement of the oxide reliability. This means that

2.4 Oxidation Parameters

HCL and Cl_2 additions provide benefits to the resulting device structures such as better ion passivation, higher and more uniform oxide dielectric strength, and improved junction properties due to lower current leakage.

The mixed flows were investigated among others by Deal and Hess in the late 70's, especially for the influence on the growth rate. The addition of H_2O as well as Cl is investigated in [32], and of HCL in [33]. In order to see the effect of the different mixed flows on the growth rate in a clear manner, the oxide thickness over time for a (100) oriented Silicon at 1000 °C is plotted in Figs. 2.7–2.9. It is notable that a double logarithmic scale of the plots leads to nearly linear curves also for the mixtures.

The mixture of H_2O/O_2 has the highest increase of the growth rate, because it is in principle a combination of wet and dry oxidation. We can see in Fig. 2.7 that the same percentage of H_2O leads to a much thicker oxide at any time than HCL or Cl_2. Another interesting aspect is that the admixture of the same percentage of HCL and Cl_2 always leads to the same oxide thickness (compare Fig. 2.8 with Fig. 2.9).

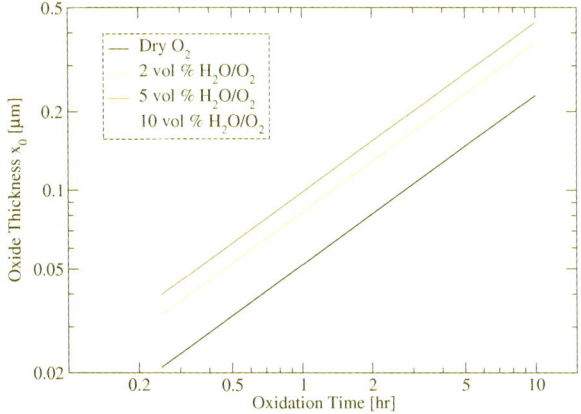

Figure 2.7: Oxide thickness versus oxidation time for (100) oriented silicon in various H_2O/O_2 mixtures at 1000 °C.

The chemical reaction of HCL with oxygen is

$$4\,\mathrm{HCL} + O_2 \rightarrow 2\,H_2O + 2\,Cl_2. \tag{2.3}$$

Now it can be said that 2 moles of HCL produce 1 mol of H_2O and Cl_2. So the mixtures of HCL can be compared with H_2O. From the theoretical aspect the double percentage of HCL should lead to the same growth effect as the single percentage of H_2O. But in the practical experiment, as shown in Fig. 2.10, 5 vol% H_2O results in a considerable thicker

PHYSICS OF THERMAL OXIDATION

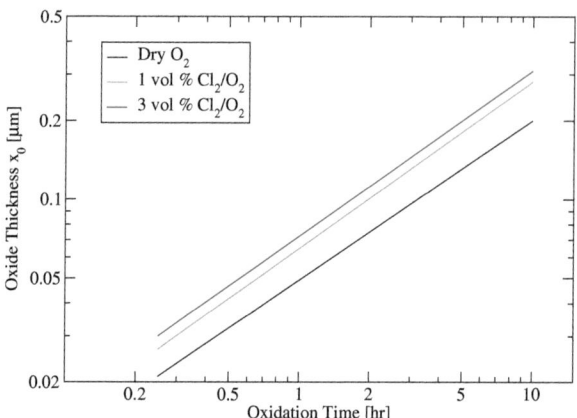

Figure 2.8: Oxide thickness versus oxidation time for (100) oriented silicon in various Cl_2/O_2 mixtures at 1000 °C.

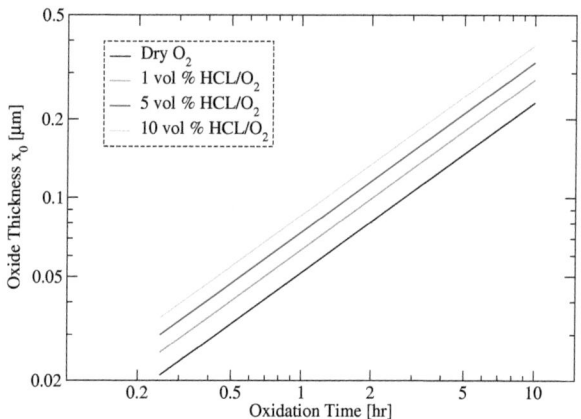

Figure 2.9: Oxide thickness versus oxidation time for (100) oriented silicon in various HCL/O_2 mixtures at 1000 °C.

oxide than 10 vol% HCL. There are no more details known about this fact [32], only that the difference between the oxide thicknesses by H_2O and HCL becomes smaller with increasing temperature, so that the theory comes true for high temperatures (1100 °C).

In wet oxidation the addition of HCL does not increase the oxidation rate, rather the oxidation rate is decreased for the same percentage as the amount of HCL is added [34].

2.4 Oxidation Parameters

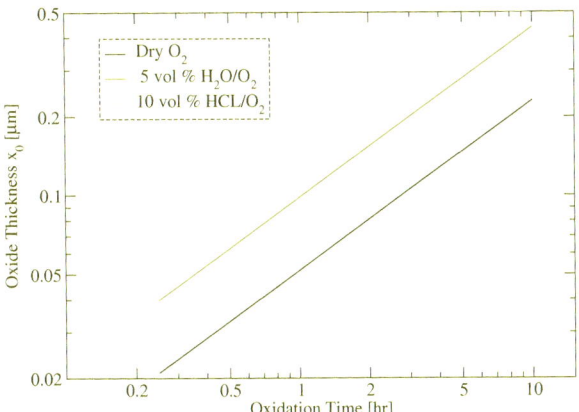

Figure 2.10: Oxidation rate of H_2O/O_2 mixture compared with HCL/O_2 mixture at 1000 °C.

In H_2O-HCL ambients the thickness uniformity and appearance of these oxides were considerably better than in pure H_2O ambients. Also the defects in the oxide are considerably reduced.

2.4.2 Influence of Temperature

The oxidation rate increases significantly with the temperature in the furnace for dry as well as for wet oxidation. The temperature dependence of the oxidation rate is plotted in Fig. 2.11 for dry and Fig. 2.12 for wet oxidation. For wet oxidation in Fig. 2.12 it can be seen that 100 °C more temperature leads to approximately double the oxidation rate, if the temperature is increased from 900 to 1000 °C. The important temperature effect can also be observed for dry oxidation in Fig. 2.11, where the same temperature increase from 900 to 1000 °C leads to much more than double the oxidation rate.

The main reason of this striking temperature influence on the oxidation rate is the temperature dependence of the diffusivity of oxygen (O_2) and water (H_2O) in fused silica. The diffusivity of the oxidants depends on the temperature T in the way $\exp(-\frac{c}{T})$. The oxidant diffusivity is exponentially increased with higher temperature and exponentially decreased with lower temperature. Higher diffusivity means that more oxidants can reach the Si/SiO_2 interface and react there with silicon to form SiO_2.

PHYSICS OF THERMAL OXIDATION

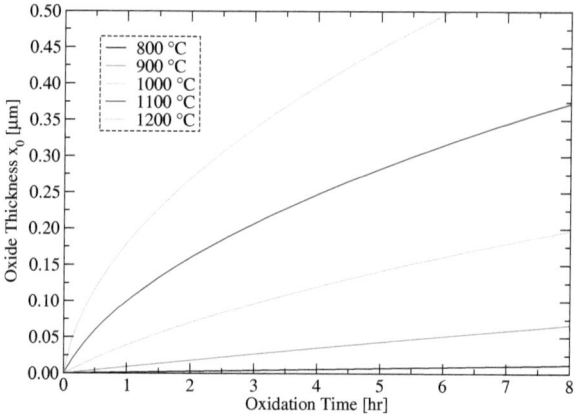

Figure 2.11: Oxide thickness versus oxidation time for (100) oriented silicon by dry oxidation (O_2) for various temperatures.

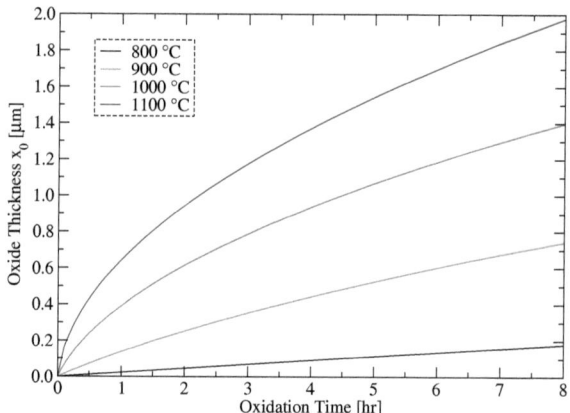

Figure 2.12: Oxide thickness versus oxidation time for (100) oriented silicon by wet oxidation (H_2O) for various temperatures.

2.4 Oxidation Parameters

2.4.3 Influence of Pressure

The oxidation rate increases with the hydrostatic pressure in the furnace for dry and wet oxidation in nearly the same way. The principal advantages of higher pressure oxidation over conventional atmospheric oxidation are the faster oxidation rate (see Fig. 2.13) and the lower processing temperature generally employed [35, 36]. Both lead to less impurity diffusion and minimum junction movement during the several oxidation steps which are necessary in the manufacturing of high-density multilayer IC devices. The quality and integrity of higher pressure oxides have been found to be comparable to atmospheric oxides. Oxidation-induced stacking faults are significantly reduced with higher pressure oxidation [37], which leads to improved device performance.

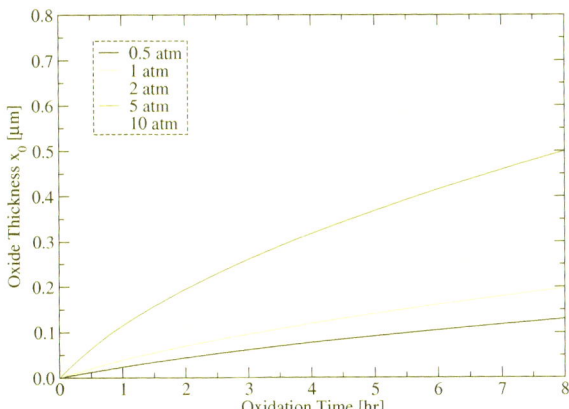

Figure 2.13: Oxide thickness versus oxidation time for (110) oriented silicon by dry oxidation at 1000 °C for various pressures.

2.4.4 Influence of Crystal Orientation

The studies of oxidation have shown that the oxidation rate also depends on the crystal orientation of the silicon substrate. Experiments have demonstrated many times that the oxide growth is faster on (111) oriented surfaces than on (100) oriented at any temperature for dry as well as wet oxidation. Furthermore, as plotted in Fig. 2.14 for wet oxidation, it was found that the (111) and (100) orientation represent the upper and the lower bound for oxidation rates, respectively. Therefore, the growth rate for all other orientations lies between these two extremal values [38].

It is important to understand orientation effects on oxidation more generally because

PHYSICS OF THERMAL OXIDATION

many structures actually use etched trenches and other shaped silicon regions as part of their structure. Ligenza [39] suggested that the crystal orientation effect might be caused by differences in the surface density of silicon atoms on the various crystal faces. He argued that since silicon atoms are required for the oxidation process, crystal planes that have higher densities of atoms should oxidize faster. Furthermore, he argued that not only the number of silicon atoms per cm^2 is important, but also the number of bonds matter, since it is necessary for Si-Si bonds to be broken for proceeding the oxidation. Ligenza calculated the "available" bonds per cm^2 on the various silicon surfaces and concluded that oxidation rates in H_2O ambients should be in the order $(111) > (100)$, which was also observed experimentally.

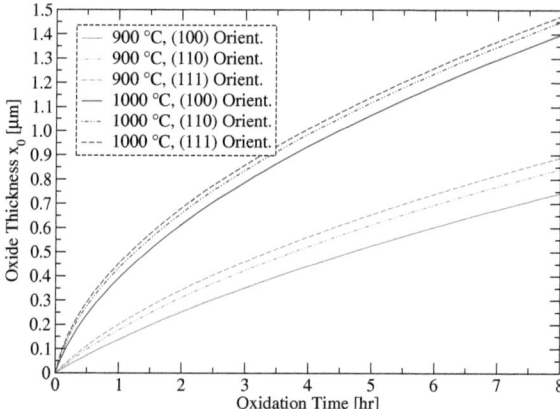

Figure 2.14: Oxide thickness versus oxidation time for (100), (110), and (111) oriented silicon by wet oxidation (H_2O).

2.5 Nitrided Oxide Films

While SiO_2 was the main material for gate dielectrics for more than three decades, the use of traditional SiO_2 gate dielectrics becomes questionable for sub-0.25 µm ULSI devices. Increasing problems with dopant penetration through ultrathin SiO_2 layers ($<2\,\text{nm}$) and direct tunneling for ultrathin oxide films dictate the search for new materials for future gate dielectrics with better diffusion barrier properties and higher dielectric constants [40]. At this time, ultrathin silicon oxynitrides (SiO_xN_y) are the leading candidates to replace pure SiO_2 [41].

Nitrogen suppresses boron penetration from the poly-Si gate and reduces hot-electron-induced degradation. The dielectric constant of the oxynitride increases linearly with the percentage of nitrogen from $\varepsilon_{SiO_2} = 3.8$ to $\varepsilon_{Si_3N_4} = 7.8$. Because most SiO_xN_y films are currently grown by thermal methods, they are only lightly doped with N (<10 at.%). Therefore, these silicon oxynitrides have a dielectric constant only slightly higher than that of pure SiO_2.

2.5.1 Different Nitridation Methods

The performance of MOS-based devices depends on both the concentration and distribution of the nitrogen atoms incorporated into the gate dielectric. The optimal nitrogen profile is determined by its application. One possibility is a SiO_xN_y film with two nitrogen-enhanced layers: at first, nitrogen is placed at or near the Si/SiO_2 interface to improve hot-electron immunity, and second, an even higher nitrogen concentration is put at the SiO_2/polysilicon interface where it is best used to minimize the penetration of boron from the heavily doped gate electrode [42]. Typical amounts of nitrogen at each interface are in the order of $(0.5-1) \times 10^{15}\,\text{cm}^{-2}$.

Nitrogen may be incorporated into SiO_2 using either thermal oxidation/annealing or chemical and physical deposition methods. Thermal nitridation of SiO_2 in NO or N_2O generally results in a relatively low concentration of nitrogen in the films in the order of 10^{15} N/cm^2 [42]. Since the nitrogen content increases with temperature, thermal oxynitridation is typically performed at high temperatures ($T > 800\,°C$).

For more heavily N-doped SiO_xN_y films, other deposition methods, such as chemical vapor deposition in different variants, or nitridation by energetic nitrogen particles (e.g. N atoms or ions), can be used. These nitridation methods can be performed at lower temperatures (~ 300–$400\,°C$). Unfortunately, low temperature deposition methods result in non-equilibrium films, and subsequent thermal processing steps are often required to improve film quality and minimize defects and induced damage [43].

2.5.2 Diffusion-Barrier Properties of Nitrided Layers

An important property of nitrogen in nitrided oxides is that it forms a barrier against the diffusion of boron. Concurrent with this, it also lowers the diffusion rates for oxygen and other dopants, slowing down the growth rate of any further oxidation or nitridation [44]. For example, for a 2 nm oxynitride with one monolayer of nitrogen 6.8×10^{14} N/cm^2 located near the interface, the oxidation rate decreases by at least a factor 4 relative to the pure oxide. The decrease in film growth rate results from a decreased rate of diffusion due to nitrogen.

One explanation for the lower diffusivity of NO, O_2, N_2 or other molecular species is the higher density of nitrides and oxynitrides compared with pure oxide. Furthermore, the lattice involves N bonds and therefore becomes more rigid. The three bonds connected to each nitrogen as in Si_3N_4 are more constrained than the two bonds of each O atom in SiO_2, where the Si–O–Si bond angles can go from 120° to 180° with little change in energy. These more constrained bonds are another important reason for decreasing the ability of nitrided lattices to permit the diffusion of atoms and small molecules.

2.5.3 Nitrogen Incorporation by NO

Oxidation of silicon and annealing of SiO_2 in nitric (NO) or nitrous (NO_2) oxide are the leading procedures for making nitrided oxides by conventional thermal processing methods. NO is the main species responsible for nitrogen incorporation into the film [45]. Oxynitridation in pure NO should be considered for ultrathin dielectrics, especially in processes where thermal budget and film thickness issues are crucial. When the temperature increases, the total amounts of both nitrogen and oxygen increase as well as the ratio of nitrogen to oxygen so that the film becomes more nitride-like at higher temperatures. For example the ratio increases by 40% if the temperature changes from 700 to 1000 °C [46]. With rising temperature the depth of the nitrogen profiles and so the width of the containing nitrogen region increase too.

The thicknesses of the films on clean silicon surfaces measured at 700–1000 °C after one hour were only $\sim$1.5–2.5 nm [46]. From the practical point of view, the slower growth of oxynitride compared with pure oxide facilitates good thickness control in the ultrathin regime during high-temperature processing. To make a thicker film, a thin preoxide (SiO_2) of desired thickness can first be formed and then annealed by NO. However, the nitrogen distribution in NO-annealed films is different compared to the one in NO-grown filmss (see Fig. 2.15).

2.5.4 Nitrogen Incorporation and Removal by NO_2

Under equivalent conditions, oxynitridation in NO_2 results in less nitrogen incorporation than in NO. However, NO_2 is particularly attractive, because

1) it allows to incorporate an appropriate amount of nitrogen near the SiO_xN_y/Si interface (typically $\sim 5 \times 10^{14}$ atoms/cm^2),
2) its processing with O_2 gas permits NO_2 to replace oxygen in the oxidation reactors and furnaces respectively.

Among other factors, oxynitridation in NO_2 is complicated by the fast gas-phase decomposition of the molecule into N_2, O_2, NO, and O at typical oxidation temperatures 800–1000 °C [47], in contrast to NO, which is a relatively stable molecule.

The fundamental difference between oxynitridation in NO_2 and NO is that, while both incorporate nitrogen by NO reactions near the interface, in the NO_2 case the nitrogen incorporation occurs simultaneously with nitrogen removal from the upper layers of the film (see Fig. 2.15). In experiments it was observed that NO does not effectively remove nitrogen from the oxynitride [48]. So it can be concluded that other products of the NO_2 gas-phase decomposition, like O, are responsible for the nitrogen removal. The

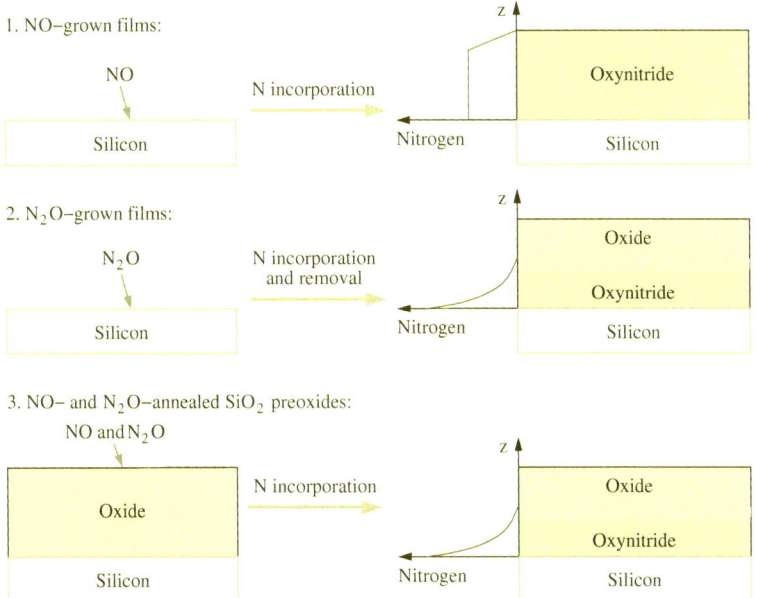

Figure 2.15: Several nitridation processes and resultant nitrogen profiles.

PHYSICS OF THERMAL OXIDATION

final nitrogen concentration and distribution is influenced by a competition between N incorporation and removal.

NO_2 rapidly decomposes in the gas phase to N_2 and O, and then the O initiates a further series of reaction to form NO, the key oxynitriding agent, and other species. NO, from gas or decomposition, is similar to O_2 when it interacts with silicon or SiO_2, in that the dominant oxynitride growth mechanism involves NO diffusion through a SiO_xN_y overlayer, followed by a reaction with silicon at and near the SiO_xN_y/Si interface [48].

2.5.5 Nitridation in N_2 and NH_3

Direct nitridation via reaction of silicon with N_2 requires very high temperatures (T $\geq 1200\,°C$) and, therefore, a too high thermal budget. To reduce the thermal budget, oxynitrides were grown in pure N_2 by rapid thermal processing (RTP). Although the input N_2 gas stream is purified at the point of use and therefore extremely free of contaminants such as N_2O, O_2, CO_2, and CO (less than 1 ppb each of them), it was found in experiments, that in a cold wall RTP module, the growth chamber contributes impurities to the ambient through outgasing from the walls [49].

Therefore, although the Si/N_2 system may be inert for $T \leq 1200\,°C$, the de facto oxidation ambient is not so. Thus, it was observed that N_2 reacts with silicon at moderate temperatures (760–1050 °C) in an RTP module [49], due to the presence of gas impurities, to form ultrathin (less than 1.2 nm) SiO_xN_y films.

Nitridation in ammonia (NH_3) was one of the first methods used to incorporate relatively high concentrations of nitrogen ($\sim 10-15$ at.%) into SiO_2 films. The nitridation atmosphere of NH_3 introduces high concentrations of hydrogen into SiO_2 films, which then can act as traps. One of the advantages of the thermal nitridation of SiO_2 in NH_3 is the simultaneous nitridation of the interface and the SiO_2 surface, while one disadvantage is the introduction of hydrogen in the oxynitride film. This disadvantage can be overcome by performing a thermal reoxidation of the oxynitride film in dry O_2, which completely removes the hydrogen from the film and serves also to decrease the concentration of nitrogen at the SiO_2/Si interface, improving the electrical characteristics of this interface [50].

The role of hydrogen is crucial, because, if hydrogen is not contained in the nitriding molecule as in the case of thermal treatments of SiO_2 films in N_2 or N_2/H_2 mixtures, incorporation of nitrogen in the films does not occur. Hydrogen participates in the transport of the nitriding species from the film surface towards the SiO_2/Si interface. Ammonia reacts in the surface region of silica at temperatures above 650 °C as

$$Si - O - Si + NH_3 \rightarrow SiNH_2 + SiOH. \qquad (2.4)$$

At the SiO_2/Si interface, because of the existence of free silicon atoms and by considering the change of free energy of the chemical reactions, the following reactions can take place

$$\mathrm{SiO_2 + 3\,Si + 4\,NH_3 \rightarrow 2\,Si_2N_2O + 6\,H_2}. \tag{2.5}$$

$$\mathrm{2\,SiO_2 + Si + 4\,NH_3 \rightarrow Si_3N_4 + 4\,H_2O + 6\,H_2}. \tag{2.6}$$

In the bulk of the silicon oxide, on the other side, the nitriding species will react mostly with silicon oxide

$$\mathrm{2\,SiO_2 + Si + 4\,NH_3 \rightarrow Si_3N_4 + 4\,H_2O + 6\,H_2}. \tag{2.7}$$

The incorporation of a nitrogen atom will often be accompanied by the intake of a hydrogen atom which removes an oxygen atom (in form of water or OH), the nitridation of the SiO_2 films proceeds essentially by an exchange of N for O atoms [50].

2.6 The Deal-Grove Model

A well established model for thermal oxide growth has been proposed by Deal and Grove [51] in the middle of the 60's and because of its simplicity it is still applied frequently. One reason for this simplicity is that the whole physics of the oxidation process is contained in two so-called Deal-Grove parameters, which must be extracted from experiments. Furthermore, it is assumed that the structure is one-dimensional. Therefore, the model can only be applied to oxide films grown on plane substrates.

2.6.1 Concept and Formulation

If one assumes that the oxidation process is dominated by the inward movement of the oxidant species, the transported species must go through the following stages:

(1) It is transported from the bulk of the oxidizing gas to the outer surface of oxide, where it is adsorbed.

(2) It is transported across the oxide film towards silicon.

(3) It reacts at the interface with silicon and form a new layer of SiO_2.

Each of these steps can be described as independent flux equation. The adsorption of oxidants is written as

$$F_1 = h(C^* - C_O), \tag{2.8}$$

where h is the gas-phase transport coefficient, C^* is the equilibrium concentration of the oxidants in the surrounding gas atmosphere, and C_O is the concentration of oxidants at the oxide surface at any given time.

PHYSICS OF THERMAL OXIDATION

It was found experimentally that wide changes in gas flow rates in the oxidation furnaces, changes in the spacing between wafers on the carrier in the furnace, and a change in wafer orientation (standing up or lying down) cause only little difference in oxidation rates. These results imply that h is very large, or that only a small difference between C^* and C_O is required to provide the necessary oxidant flux.

C^* is also the solubility limit in the oxide, which is assumed to be related to the partial pressure p of the oxidant in the gas atmosphere by Henry's law

$$C^* = H \cdot p. \tag{2.9}$$

At natural ambient pressure of 1 atm and at a temperature of 1000 °C, the solubility limits are 5.2×10^{16} cm^{-3} for O_2, and 3.0×10^{19} cm^{-3} for H_2O.

The flux F_2 represents the diffusion of the oxidants through the oxide layer to the Si-SiO$_2$-interface, which can be expressed as

$$F_2 = D\frac{\partial C}{\partial x} = D\frac{C_O - C_S}{x_O}, \tag{2.10}$$

where D is the oxidant diffusivity in the oxide, C_S is the oxidant concentration at the oxide-silicon interface, and x_O represents the oxide thickness. In this expression it is assumed that the process is in steady state (no changing rapidly with time), and that there is no loss of oxidants when they diffuse through the oxide. Under these conditions, F_2 must be constant through the oxide and hence the derivative can be replaced simply by a constant gradient.

The third part of the oxidation process is the flux of oxidants consumed by the oxidation reaction at the oxide-silicon interface given by

$$F_3 = k_s C_S, \tag{2.11}$$

with k_s as the surface rate constant. k_s really represents a number of processes occurring at the Si/SiO$_2$ interface. These may include oxidant ($O_2 \to 2O$), Si-Si bond breaking, and/or Si-O bond formation. The rate at which this reaction takes place should be proportional to the oxidant concentration at the interface C_S.

Deal and Grove assumed that in the steady state condition these three fluxes are equal, which allows to express them as

$$F_1 = F_2 = F_3 = F = \frac{C^*}{\frac{1}{k_s} + \frac{1}{h} + \frac{x_O}{D_0}}. \tag{2.12}$$

The rate of oxide growth is proportional to the flux of oxidant molecules,

$$\frac{dx_0}{dt} = \frac{F}{N} = \frac{\frac{C^*}{N}}{\frac{1}{k_s} + \frac{1}{h} + \frac{x_0}{D_0}}, \tag{2.13}$$

2.6 The Deal-Grove Model

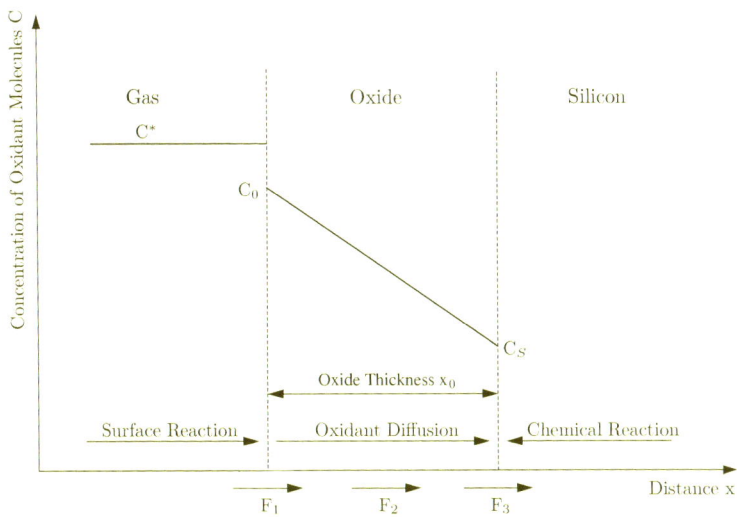

Figure 2.16: One-dimensional model for the oxidation of silicon.

where N is the number of oxidant molecules incorporated per unit volume.

The differential equation can be simplified as

$$\frac{dx_0}{dt} = \frac{B}{A + 2x_0}, \tag{2.14}$$

with the physically based parameters

$$A = 2D\left(\frac{1}{k_s} + \frac{1}{h}\right), \tag{2.15}$$

$$B = 2D\frac{C^*}{N}. \tag{2.16}$$

2.6.2 Analytical Oxidation Relationship

In order to get an analytical relationship between oxide thickness x_0 and oxidation time t the first order differential equation (2.14) must be solved. For this purpose in the first step (2.14) can be rewritten in the form

$$(A + 2x_0)\, dx_0 = B\, dt. \tag{2.17}$$

Integration of (2.17) from time 0 to t, with the assumption of an initial oxide thickness x_i at time 0, yields a quadratic equation for the oxide thickness x_0:

$$x_0^2 + Ax_0 = B(t + \tau), \tag{2.18}$$

PHYSICS OF THERMAL OXIDATION

where the parameter τ is given by

$$\tau = \frac{x_i^2 + Ax_i}{B}. \tag{2.19}$$

So τ takes into account any oxide thickness at the start of the oxidation. It can also be used to provide a better fit to the data in the anomalous thin oxide regime in dry oxidation.

At first with (2.18) the oxidation time for a specific desired oxide thickness can be estimated by

$$t = \frac{x_0^2 - x_i^2}{B} + \frac{x_0 - x_i}{B/A}. \tag{2.20}$$

On the other side solving the quadratic equation (2.18) in regard of x_0 leads to the following explicit expression for the oxide thickness in terms of oxidation time:

$$x_0 = \frac{A}{2}\left(\sqrt{1 + \frac{4B}{A^2}(t + \tau)} - 1\right). \tag{2.21}$$

The formulas (2.44) and (2.20) are a real strength of the Deal-Grove model, because the oxide thickness for any oxidation time or the needed time for a specific thickness can be determined in an uncomplicated and fast way. Of course the thickness can be only estimated in one direction on planar structures, but in practice this fast approach is indeed helpful.

It is interesting to examine two limiting forms of the linear-parabolic relationship (2.44). One limiting case occurs for long oxidation times when $t \gg \tau$ and $t \gg A^2/4B$

$$x_0 \cong \sqrt{B \cdot t}, \tag{2.22}$$

where B is the so-called parabolic rate constant

$$B = \frac{2DC^*}{N}. \tag{2.23}$$

The other limiting case occurs for short oxidation times when $t \ll A^2/4B$

$$x_0 \cong \frac{B}{A}(t + \tau), \tag{2.24}$$

where B/A is the so-called linear rate constant

$$\frac{B}{A} = \frac{C^*}{N\left(\frac{1}{k_s} + \frac{1}{h}\right)} \cong \frac{C^* k_s}{N}. \tag{2.25}$$

The linear term (2.24) dominates for small x-values, the parabolic term (2.22) for larger x-values.

The rate constants B and B/A are also termed as Deal-Grove-parameters. In most publications which use the Deal-Grove model the oxide growth is described with B and B/A. The parameters B and B/A are normally determined experimentally by extracting them from growth data. The reason for taking this approach is simply that all parameters in (2.23) and (2.25) are not known. k_s in particular contains a lot of hidden physics associated with the interface reaction.

2.6.3 Temperature Dependence of B and B/A

In order to model the corresponding growth rate for different temperatures, the values for B and B/A must change with temperature. As explained in Section 2.4.2, the oxidation rate increases with higher temperature, and so the values of B and B/A must also increase. It was found experimentally that both B and B/A are well described by Arrhenius expressions of the form

$$B = C_1 \exp\left(-\frac{E_1}{kT}\right) \tag{2.26}$$

$$\frac{B}{A} = C_2 \exp\left(-\frac{E_2}{kT}\right). \tag{2.27}$$

In these expressions, E_1 and E_2 are the activation energies associated with the physical process that B and B/A represents, and C_1 and C_2 are the pre-exponential constants. Table 2.2 lists the experimental values for the parameters needed in (2.26) and (2.27) for (111) oriented silicon at one atmosphere. With these values, in Fig. 2.17 the parameters B and B/A are plotted over the temperature range 800 - 1000 °C for wet and dry oxidation. In order to get the corresponding values for (100) oriented silicon, only the C_2 values must be divided by the factor 1.68, all the $E_{1,2}$ and C_1 values are the same.

Table 2.2: Arrhenius parameters for B and B/A in (111) oriented silicon [25].

Ambient	B	B/A
Dry O_2	$C_1 = 7.72 \times 10^2 \mu m^2/hr$	$C_2 = 6.23 \times 10^6 \mu m/hr$
	$E_1 = 1.23$ eV	$E_2 = 2.00$ eV
Wet H_2O	$C_1 = 3.86 \times 10^2 \mu m^2/hr$	$C_2 = 1.63 \times 10^8 \mu m/hr$
	$E_1 = 0.78$ eV	$E_2 = 2.05$ eV

For the parabolic rate constant B the activation energy E_1 is quite different for O_2 and H_2O ambients. (2.23) suggests that the physical mechanism responsible for E_1 might be the oxidant diffusion through SiO_2, because N is a constant and C^* is not expected to increase exponentially with temperature. In fact, independent measurements of the diffusion coefficients of O_2 and H_2O in SiO_2 show that these parameters vary with temperature

PHYSICS OF THERMAL OXIDATION

in the same way as (2.26) and with E_1 values close to those shown in Table 2.2. The clear implication is that B in the linear parabolic model really represents the oxidant diffusion process.

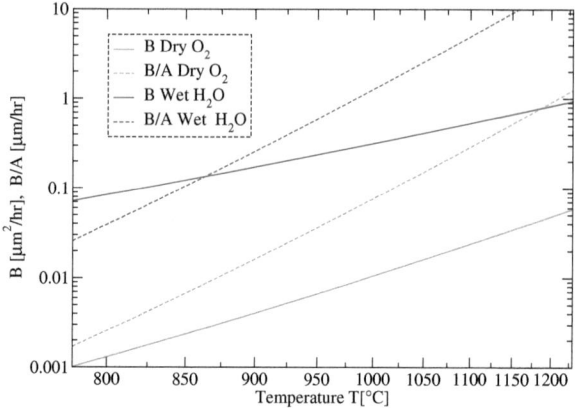

Figure 2.17: B and B/A versus temperature for (111) oriented silicon for wet and dry oxidation.

The E_2 values for B/A in the table are all quite close to 2 eV. (2.25) suggests that the physical origin of E_2 is likely connected with the interface reaction rate k_s. Traditionally, the 2 eV activation energy has been associated with the Si-O bound formation process because of measurements by Pauling [52] that suggested that the Si-O bond energy was in the correct range to explain the B/A values. However, the interface reaction is very complex and it is likely that other effects also affect the experimental B/A values. An additional observation supports the idea that it is somehow associated with the silicon substrate which determines E_2, because E_2 is essentially independent of the oxidation ambient. It is also essentially independent of the substrate crystal orientation, which suggests that E_2 represents a fundamental part of the oxidation process, not something only associated with the substrate.

2.6.4 Pressure Dependence of B and B/A

The linear parabolic model predicts that the oxide growth rate should be directly proportional to the oxidant pressure as shown in (2.9). If Henry's law [53] holds and the concentration of oxidants on the gas/SiO_2 interface C^* is proportional to the pressure p, then both B and B/A are proportional to p from (2.23) and (2.25), and the oxide growth rate should therefore be proportional to p.

2.6 The Deal-Grove Model

Experimental measurements have shown that for wet oxidation this prediction is correct, and for H_2O ambients the pressure dependence of the parabolic and linear rate constants are [36]

$$B(P) = B(1\text{atm}) \cdot p, \tag{2.28}$$

$$B/A(P) = B/A(1\text{atm}) \cdot p. \tag{2.29}$$

In contrast to wet oxidation for dry oxidation the pressure dependence is inconsistent with the linear parabolic model. A considerable body of data has consistently shown that dry oxidation can only be modeled with a linear parabolic equation, where $B \propto p$ and $B/A \propto p^n$ with n $\approx$ 0.7 − 0.8 [35]. Hence, to use the model for O_2 ambients at any pressure p the parabolic and linear rate constants should be

$$B(P) = B(1\text{atm}) \cdot p \tag{2.30}$$

$$B/A(P) = B/A(1\text{atm}) \cdot p^{0.75}. \tag{2.31}$$

Within the context of the model it can be inferred that the pressure dependence of $B \propto p$ comes exclusively from C^*, because C^* as determined in Henry's law (2.23) must be $C^* \propto p$. Therefore, the diffusion coefficient D for the oxidants in the solid phase can be assumed constant.

If B/A is not linearly proportional to p, k_s from (2.25) must depend on p in a non-linear fashion. Considering the pressure dependence of B/A and C^* above, the chemical surface reaction must depend on pressure in the way $k_s \propto p^{-0.25}$.

2.6.5 Dependence of B and B/A on Crystal Orientation

Even before the development of the Deal-Grove model, it has been observed that crystal orientation affects the oxidation rate [39]. The crystal effects can be incorporated in the following way: Except perhaps in the region very near the Si/SiO_2 interface, the oxide grows on silicon in an amorphous way. So it does not incorporate any information about the underlying silicon crystal structure. Therefore, the parabolic rate constant B should not be orientation dependent, since B represents the oxidant diffusion through the SiO_2. If the oxide structure is unrelated to the underlying substrate, there should be no crystal orientation effect on B. In fact it was found experimentally by extracting growth data [38], that in context of the model there is no crystal effect on the rate constant B. The B values are the same for all orientations.

On the other hand B/A should be orientation dependent, because it involves the reaction at the Si/SiO_2 interface. This reaction surely involves silicon atoms and should be affected by the number of available reaction sites. It was found experimentally [38], that there are two extremes of the linear rate constant B/A. The minimum was found for (100) oriented silicon whereas the maximum is at (111) orientation, and all other orientation

PHYSICS OF THERMAL OXIDATION

are normally between these two extremes. In the context of the model the orientation effect must be incorporated for the rate constant B/A in the following way [38]:

$$B/A\langle 111\rangle = 1.68 \cdot B/A\langle 100\rangle. \tag{2.32}$$

2.6.6 Thin Film Oxidation with Deal-Grove Model

It has been observed in many experiments that there is a rapid and non-linear oxide growth in the initial stage of dry oxidation [54], as presented in Fig. 2.18. One weakness of the model is the impossibility to predict the initial stage of the oxidation growth. As shown in Fig. 2.18, even with the best fit, the approximately first 30 nm of the oxide thickness can not be forecasted with the linear parabolic model, because the oxide growth is fast and non-linear but the model offers only a linear fit for such thin thicknesses [55].

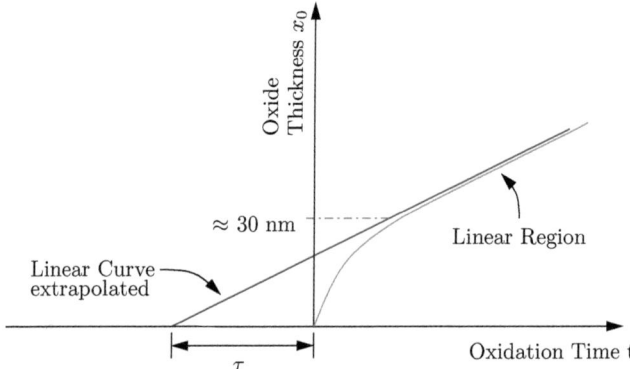

Figure 2.18: Rapid, non-linear growth rate in the inital stage of dry oxidation.

2.7 The Massoud Model

As described in the Section 2.6.6 the Deal-Grove model can not satisfy the so-called thin film oxidation. It should be taken into account that in the middle of the 60's, when Deal and Grove developed their model, oxide thicknesses under 30 nm were not fabricated in the semiconductor technology. Hence, there was no need to predict or simulate the growth for such thin oxide films. But with shrinking device geometry also the oxide thickness is decreasing. Hence, sometime in the 80's, MOS gates with thin thicknesses were grown and so the problem became important. In order to handle also thin film oxditation, in this time Massoud and other people, reengineered the Deal-Grove concept [56]. The yielded model

2.7.1 Experimental Fitting

It was found that the SiO$_2$ growth rate in the thin regime for a wide variety of experimental conditions can be expressed as [56]

$$\frac{dx_0}{dt} = \frac{B}{2x_0 + A} + C_1 \exp\left(-\frac{x_0}{L_1}\right) + C_2 \exp\left(-\frac{x_0}{L_2}\right). \quad (2.33)$$

The first term on the right side of (2.33) is the linear-parabolic term where B and B/A are the parabolic and linear rate constants, respectively, as defined by Deal and Grove, but their values in the Massoud model are completely different [57]. In Arrhenius-expression the rate constants can be written in the form

$$B = C_B \exp\left(-\frac{E_B}{kT}\right) \quad (2.34)$$

$$\frac{B}{A} = C_{B/A} \exp\left(-\frac{E_{B/A}}{kT}\right). \quad (2.35)$$

The values for the pre-exponential constants C_B, $C_{B/A}$ and the activation energies E_B, $E_{B/A}$ for different crystal orientations are listed in Table 2.3. In this model it is adverse that C_x and E_x are not valid for the whole temperature range and so C_x and E_x differ from temperatures less and more than 1000 °C.

Table 2.3: Pre-exponential constants and activation energies for B and B/A [57].

Temperature Range	T < 1000 °C			T > 1000 °C	
Crystal Orientation	(100)	(111)	(110)	(100)	(111)
C_B [nm^2/min]	1.70×10^{11}	1.34×10^9	3.73×10^8	1.31×10^5	2.56×10^5
E_B [eV]	2.22	1.71	1.63	0.68	0.76
$C_{B/A}$ [nm/min]	7.35×10^6	1.32×10^7	4.73×10^8	3.53×10^{12}	6.50×10^{11}
$E_{B/A}$ [eV]	1.76	1.74	2.10	3.20	2.95

In (2.33) the two exponential terms represent the rate enhancement in the thin regime. They are defined in terms of pre-exponential constants C_1 and C_2 and characteristic lengths L_1 and L_2. The first decaying exponential has a characteristic lengths L_1 in the order of 1 nm, it is nonzero for the first 5 nm of oxide growth, and vanishes for oxides thicker than 5 nm. The second decaying exponential has a characteristic lengths L_2 in the order of 7 nm and it is present from the onset of oxidation to an oxide thickness of about 25 nm, where it decays to zero and the growth becomes pure linear-parabolic.

PHYSICS OF THERMAL OXIDATION

Another formulation of (2.33), where the two terms which represent the rate enhancement in the thin regime are decaying exponentially with time, can be expressed as [58]

$$\frac{dx_0}{dt} = \frac{B + K_1 \exp\left(\frac{-t}{\tau_1}\right) + K_2 \exp\left(\frac{-t}{\tau_2}\right)}{2x_0 + A}, \tag{2.36}$$

where all four parameters K_1, K_2, τ_1 and τ_2 were fitted to an Arrhenius-type expression

$$K_1 = K_1^0 \exp\left(-\frac{E_{K1}}{kT}\right), \tag{2.37}$$

$$K_2 = K_2^0 \exp\left(-\frac{E_{K2}}{kT}\right), \tag{2.38}$$

$$\tau_1 = \tau_1^0 \exp\left(-\frac{E_{\tau 1}}{kT}\right), \tag{2.39}$$

$$\tau_2 = \tau_2^0 \exp\left(-\frac{E_{\tau 2}}{kT}\right). \tag{2.40}$$

The pre-exponential constants and activation energies in the above expressions (2.37)–(2.40) for different crystal orientations and dry oxidation in the temperature range form 800–1000 °C are listed in Table 2.4.

Table 2.4: Arrhenius-expression parameters for the pre-exponential constants K_1 and K_2, and the time constants τ_1 and τ_2 in the 800–1000 °C range [58].

Crystal Orientation	(100)	(111)	(110)
K_1^0 [nm²/min]	2.49×10^{11}	2.70×10^9	4.07×10^8
E_{K1} [eV]	2.18	1.74	1.54
K_2^0 [nm²/min]	3.72×10^{11}	1.33×10^9	1.20×10^8
E_{K2} [eV]	2.28	1.76	1.56
τ_1^0 [min]	4.14×10^{-6}	1.72×10^{-6}	5.38×10^{-9}
$E_{\tau 1}$ [eV]	1.38	1.45	2.02
τ_1^0 [min]	2.71×10^{-7}	1.56×10^{-7}	1.63×10^{-8}
$E_{\tau 2}$ [eV]	1.88	1.90	2.12

2.7.2 Analytical Oxidation Relationship

As already mentioned in Section 2.6.2, it would be convenient to have an analytical expression for the oxide thickness x_0. For this purpose (2.36) is rewritten as

$$(2x_0 + A)dx_0 = \left[B + K_1 \exp\left(-\frac{t}{\tau_1}\right) + K_2 \exp\left(-\frac{t}{\tau_2}\right)\right]dt. \tag{2.41}$$

2.7 The Massoud Model

Integration of (2.41), from time 0 where the native oxide thickness is x_i to an oxidation time t where the oxide thickness x_0 results in [58]

$$x_0^2 + Ax_0 = Bt + M_1\left[1 - \exp\left(-\frac{t}{\tau_1}\right)\right] + M_2\left[1 - \exp\left(-\frac{t}{\tau_2}\right)\right] + M_0 \quad (2.42)$$

with the substitutions

$$M_0 = (x_i^2 + Ax_i), \qquad M_1 = K_1\tau_1, \qquad M_2 = K_2\tau_2. \quad (2.43)$$

The equation (2.42) is quadratic and can be solved obtaining an analytic expression for the oxide thickness as a function of the oxidation time of the form

$$x_0 = \sqrt{\left(\frac{A}{2}\right)^2 + Bt + M_1\left[1 - \exp\left(-\frac{t}{\tau_1}\right)\right] + M_2\left[1 - \exp\left(-\frac{t}{\tau_2}\right)\right] + M_0} - \frac{A}{2} \quad (2.44)$$

This relationship describes the oxide growth in dry oxygen from the onset oxidation with an smaller than 1–2% compared to the measured data. An example in Fig. 2.19 shows the growth in thin regime for (100) oriented silicon in temperature range 900–1000 °C. The oxide thicknesses were calculated by (2.42) with the parameters from Table 2.3 and 2.4.

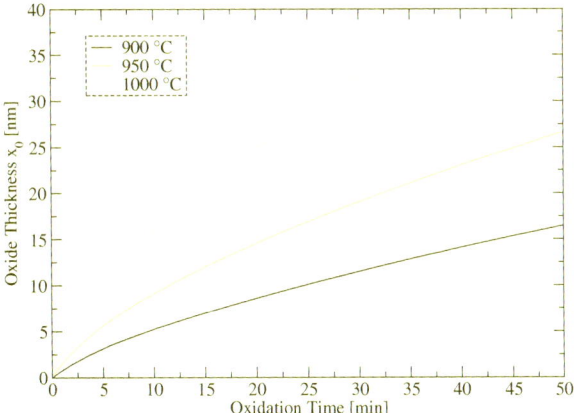

Figure 2.19: Oxide thickness versus oxidation time for (110) oriented silicon in dry oxygen at 900 °C, 950 °C, and 1000 °C.

Chapter 3

Advanced Oxidation Model

THE MODEL described in this chapter is designed for a realistic physical and three-dimensional simulation of thermal oxidation. Advantageously, this model takes into account that the diffusion of oxidants, the chemical reaction, and the volume increase occur simultaneously. Furthermore, this model does not use moving Si/SiO_2-interfaces for the SiO_2-growth like the standard models [59, 60], which are all based on the Deal-Grove model. The handling of moving interfaces problems in complex three-dimensional structures becomes very complicated and causes an enormous data update which are the most restricting factors for such applications [61, 62].

In case of oxidation there exist two segments, one for silicon and one for SiO_2, with an interface. It is not a problem to make a mesh for such structures, but the SiO_2-growth results in a moving boundary problem, which means that the interface should move after each simulation step. In order to reach the new position of the interface, new grid points are inserted and a remeshing step has to be performed [61, 63]. These mesh operations demand complicated algorithms.

The basic idea of this model is to define the regions of and SiO_2 on a single and static mesh with a separating parameter η. In this model η plays a key role, because the main interest of oxidation simulation is to predict the shape of the SiO_2-domain. Since the newly formed SiO_2 leads to a significant volume increase and so to large displacements or stresses, the modeling of the mechanics also plays an important role. Besides the oxidant diffusion and the change of η, the mechanics is an important part of the mathematical formulation.

ADVANCED OXIDATION MODEL

3.1 The Diffuse Interface Concept

The diffuse interface concept avoids a moving interface problem, because there is not a sharp interface between silicon and SiO_2 in contrast to the standard models [64, 65]. Because of the missing sharp interface there different segments for silicon and SiO_2 do not exist. In order to determine where is silicon and where is SiO_2, a parameter named normalized silicon is defined [66]

$$\eta(\vec{x}, t) = \frac{C_{Si}(\vec{x}, t)}{C_{0,Si}}. \tag{3.1}$$

Here $C_{Si}(\vec{x}, t)$ is the silicon concentration at time t and point $\vec{x}$ (x, y, z) and $C_{0,Si}$ is the concentration in pure silicon. η is 1 in pure silicon and 0 in pure silicon dioxide.

Instead of a sharp interface there is a so-called reaction layer where the diffusion of oxidants, the chemical reaction, and the volume increase occur simultaneously. This reaction layer has a spatial finite width (see Fig. 3.1), where the values of η lie between 0 and 1 [66]. The η curve always starts with 0 near silicon and ends at 1 near oxide, as shown in Fig. 3.2. The shape of this curve is given by the calculated η distribution in the reaction layer, which depends on the parameters in the model.

3.2 Mathematical Formulation

From the mathematical point of view the whole oxidation process can be described by a coupled system of partial differential equations, one for the diffusion of oxidants through SiO_2, the second for the conversion of Si into SiO_2 at the interface, and a third for the mechanical problem of the complete oxidized structure.

3.2.1 Oxidant Diffusion

The diffusion of oxidants in the domains Ω_1, Ω_2, and Ω_3 according to Fig. 3.1 is described by

$$D(T)\,\Delta C(\vec{x}, t) = k(\eta)\,C(\vec{x}, t), \tag{3.2}$$

where $\Delta = \frac{\partial^2}{\partial x^2} + \frac{\partial^2}{\partial y^2} + \frac{\partial^2}{\partial z^2}$ is the Laplace operator, $C(\vec{x}, t)$ is the oxidant concentration in the material, and $D(T)$ is the temperature dependent low stress diffusion coefficient. The boundary conditions for the diffusion equation (3.2) are

$$C = C^* \quad \text{on} \quad \Gamma_1 \quad \text{and} \quad \frac{\partial C}{\partial n} = 0 \quad \text{on} \quad \Gamma_2, \Gamma_3, \Gamma_4, \tag{3.3}$$

where C^* is the oxidant concentration in the gas atmosphere. $\frac{\partial C}{\partial n} = 0$ is a Neumann boundary condition, which means that there does not exist an oxidant flow through these boundaries.

In (3.2) $k(\eta)$ is the strength of a spatial sink and not just a reaction coefficient at a sharp interface [67]. $k(\eta)\,C(\vec{x},t)$ defines how many particles of oxygen per unit volume are transformed in a unit time interval to oxide. $k(\eta)$ is defined to be linearly proportional to $\eta(\vec{x},t)$

$$k(\eta) = \eta(\vec{x},t)\,k_{max}, \tag{3.4}$$

where k_{max} is the maximal possible strength of the sink.

Figure 3.1: Schematic domains and boundaries.

3.2.2 Dynamics of η

Because of the chemical reaction which consumes silicon, the normalized silicon concentration η is changed. In a test volume ΔV, where is assumed that the oxidant concentration C is constant during a time interval Δt there are $k(\eta)\,C(\vec{x},t)\,\Delta V\,\Delta t$ particles of oxygen which react with $k(\eta)\,C(\vec{x},t)\,\Delta V\,\Delta t/(\lambda\,N_1)$ unit volumes of silicon. By this process the silicon concentration is reduced.
The dynamics of η can be described by [67]

$$\frac{\partial \eta(\vec{x},t)}{\partial t} = -\frac{1}{\lambda} k(\eta)\,C(\vec{x},t)/N_1, \tag{3.5}$$

where λ is the volume expansion factor ($=2.25$) for the reaction from Si to SiO$_2$, and N_1 is the number of oxidant molecules incorporated into one unit volume of SiO$_2$.

In this model the dynamics of η is equivalent with the movement of the sharp Si/SiO$_2$ interface in the standard model, because η defines the silicon and oxide areas. The only difference is that here a diffuse interface (Fig. 3.1) moves, where is a mixture of silicon and oxide.

ADVANCED OXIDATION MODEL

3.2.3 Volume Expansion of the New Oxide

Because of the much lower density of oxide compared with silicon, the conversion from Si to SiO_2 leads to a significant volume increase of the new oxide. In the advanced model the conversion is not performed instantaneously, it needs some finite time. The fraction of SiO_2 in a small volume ΔV is expressed by the η value. The new generated oxide in the reaction layer is described by the change of η. For a time period Δt the η-value and the silicon fraction decreases with

$$\Delta \eta(\vec{x}, t) = -\frac{1}{\lambda} \Delta t \, k(\eta) \, C(\vec{x}, t) / N_1. \tag{3.6}$$

The additional volume in a test volume ΔV is given by

$$V^{add} = (\lambda - 1) \, \Delta \eta(\vec{x}, t) \, \Delta V. \tag{3.7}$$

Because the maximal volume increase of the oxide is limited to 1.25 times of the volume of original silicon, V^{add} in (3.7) must be scaled with $(\lambda - 1)$.

The normalized additional volume with (3.6) and (3.7) after a time Δt is

$$V_{rel}^{add} = \frac{\lambda - 1}{\lambda} \Delta t \, k(\eta) C(\vec{x}, t) / N_1. \tag{3.8}$$

An important aspect of (3.8) is that the sum of V_{rel}^{add} over all time steps can not be more than 125%, which is the maximal volume increase of the material during oxidation.

3.2.4 Diffusion Coefficient and Reaction Layer

In contrast to the standard models, where the diffusion coefficient $D(T)$ is automatically included in the parabolic rate constant B, in the advanced model $D(T)$ must be determined separately for the specific temperature and oxidant species. The most interesting oxidant species come from dry and wet oxidation.

In general the diffusion coefficient follows the expression

$$D(T) = D_0 \exp\left(-\frac{E_D}{RT}\right), \tag{3.9}$$

where D_0 is a pre-exponential diffusion constant, E_D is the activation energy in [cal], T the temperature in [K], and R is the universal gas constant with $R = 1.987$ cal/(K·mol).

For dry oxygen ambients the results of Norton [68] can be used, who found that the activation energy for diffusion of oxygen in vitreous silica is 27 kcal and the diffusion coefficient $D(T)$ is 4.2×10^{-9} cm^2/sec ($= 0.42 \, \mu$m^2/sec) at a temperature of 950 °C. With these data it is possible to calculate D_0, and (3.9) can be written for dry oxidation in the form

$$D(T) = 2.82 \times 10^{-4} \exp\left(-\frac{27\,000}{RT}\right) \text{ cm}^2/\text{sec}. \tag{3.10}$$

3.2 Mathematical Formulation

For wet oxidation the results from Moulson and Roberts [69] are most suitable. They have investigated heated silica glass in water vapour between 600 and 1200 °C and found the temperature dependent diffusion coefficient

$$D(T) = 1.0 \times 10^{-6} \exp\left(-\frac{18\,300}{RT}\right) \quad \text{cm}^2/\text{sec.} \tag{3.11}$$

In the advanced model the reaction layer has a spatial finite width d_{React}, which can vary. This width is mainly determined by the value of k_{max}. The bigger k_{max}, the steeper the concentration decay and the thinner the reaction layer. Therefore, d_{React} is inverse proportional to k_{max} and so the width can be controlled by k_{max}. This means that a small value of k_{max} (e.g. $k_{\max} \approx 50$) leads to a wide reaction layer (e.g $d_{\text{React}} \approx 100$ nm) and a big value of k_{max} (e.g. $k_{\max} \approx 500$) leads to a small reaction layer (e.g $d_{\text{React}} \approx 10$ nm).

The layer width is an important and necessary fact, because the thickness of the reaction layer must be much smaller than the thickness of the oxidized structure or the final oxide thickness. In order to apply this model also for dry or thin film oxidation with a few nm thickness, wide reaction layers are unusable.

Another interesting aspect of this model is the value of the diffusion coefficient $D_{0,\text{React}}$ in the reaction layer. In the standard model with a sharp interface the oxidants diffuse with the same D_0 through the oxide to the Si/SiO$_2$-interface where they react. This means that in the standard model no oxidants diffuse into silicon and a normalized coefficient $D_0^{norm} = 0$ in the silicon and $D_0^{norm} = 1$ in the oxide are appropriate.

In the advanced model the oxidant diffusion must not stop at the beginning of the reaction layer, because there the oxidants are needed for the chemical reaction. On the other side the oxidant diffusion should stop at the end of the reaction layer and not continue into the silicon material. A good approach for this model is that the values of D_0^{norm} run down gradually from an approximate value of 1 near the oxide area to a value of 0 near the silicon area as schematically shown in Fig. 3.2.

Since η defines the domains of oxide, reaction layer as well as silicon, and during the oxidation process the reaction layer moves into the silicon domain, $D_{0,\text{React}}$ must be a function of η. Because the value of D_0^{norm} must be 1 in SiO$_2$, where $\eta = 0$, and 0 in Si, where $\eta = 1$ (see Fig. 3.2), the most plausible function for the diffusion coefficient in the reaction layer is

$$D_{0,\text{React}}(\vec{x}) = D_0(1 - \eta(\vec{x})). \tag{3.12}$$

Another simple but good working formulation for $D_{0,\text{React}}$ in the reaction layer was found with

$$D_{0,\text{React}}(\vec{x}) = \frac{a}{\eta(\vec{x})} \qquad \text{for} \qquad D_{0,\text{React}} \leq D_0. \tag{3.13}$$

Here a is a small constant and $D_{0,\text{React}}$ must be limited to D_0 when $\eta \to 0$.

ADVANCED OXIDATION MODEL

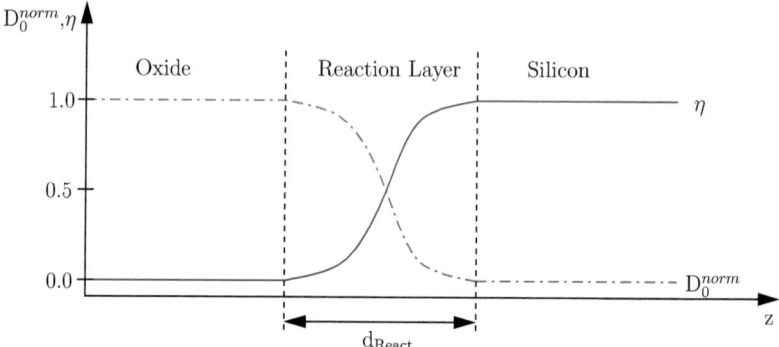

Figure 3.2: Values of η and D_0^{norm} in the reaction layer.

3.2.5 Mechanics

The chemical reaction of silicon and oxygen causes a volume increase of about 125%, which leads to significant displacements in the material. If this volume increase is only partially prevented, mechanical stress is built up in the materials. In order to calculate these displacements and stresses a mechanical modeling is needed.

In general, every three-dimensional mechanical problem can be described by the stress equilibrium relations [70]

$$\frac{\partial \sigma_{xx}}{\partial x} + \frac{\partial \sigma_{xy}}{\partial y} + \frac{\partial \sigma_{xz}}{\partial z} = f_x,$$
$$\frac{\partial \sigma_{yx}}{\partial x} + \frac{\partial \sigma_{yy}}{\partial y} + \frac{\partial \sigma_{yz}}{\partial z} = f_y, \qquad (3.14)$$
$$\frac{\partial \sigma_{zx}}{\partial x} + \frac{\partial \sigma_{zy}}{\partial y} + \frac{\partial \sigma_{zz}}{\partial z} = f_z.$$

During the oxidation process there are normally no external forces, because a volume increase caused by a chemical reaction, or a thermal expansion only lead to internal forces. Therefore, on the right-hand side of (3.14) the external forces are $f_x = f_y = f_z = 0$.

3.2.5.1 Elastic Mechanical Model

For linear elastic materials which are described by the Hook's law, the stress tensor $\tilde{\sigma}$ from (3.14) is given by

$$\tilde{\sigma} = \mathbf{D}(\tilde{\varepsilon} - \tilde{\varepsilon}_0) + \tilde{\sigma}_0. \qquad (3.15)$$

Here $\mathbf{D}$ is the so-called material matrix. Furthermore, $\tilde{\varepsilon}$ is the strain tensor, $\tilde{\varepsilon}_0$ is the residual strain tensor, and $\tilde{\sigma}_0$ is the residual stress tensor.

3.2 Mathematical Formulation

For constructing the material matrix $\mathbf{D}$, the components of the stress tensor without residual stress and strain components can be expressed in Lamé's form by [71]

$$\sigma_{ij} = \lambda\, \varepsilon_{kk}\, \delta_{ij} + 2\,\mu\, \varepsilon_{ij}, \tag{3.16}$$

where ε_{kk} is the trace of the strain tensor

$$\varepsilon_{kk} = \varepsilon_{xx} + \varepsilon_{yy} + \varepsilon_{zz}, \tag{3.17}$$

δ_{ij} is the Kronecker symbol

$$\delta_{ij} = \begin{cases} 0 \text{ for } i \neq j \\ 1 \text{ for } i = j \end{cases}, \tag{3.18}$$

and λ and μ are the so-called Lamé's constants

$$\lambda = \frac{\nu E}{(1+\nu)(1-2\nu)}, \qquad \mu = \frac{E}{2(1+\nu)}. \tag{3.19}$$

Thereby E is the Young modulus and ν is the Poisson ratio. Note, the often used shear modulus G is identical with Lamé's constant μ.

The strain tensor is

$$\tilde{\varepsilon} = \begin{bmatrix} \varepsilon_{xx} & \varepsilon_{xy} & \varepsilon_{xz} \\ \varepsilon_{yx} & \varepsilon_{yy} & \varepsilon_{yz} \\ \varepsilon_{zx} & \varepsilon_{zy} & \varepsilon_{zz} \end{bmatrix} = \begin{bmatrix} \varepsilon_{xx} & \frac{1}{2}\gamma_{xy} & \frac{1}{2}\gamma_{xz} \\ \frac{1}{2}\gamma_{yx} & \varepsilon_{yy} & \frac{1}{2}\gamma_{yz} \\ \frac{1}{2}\gamma_{zx} & \frac{1}{2}\gamma_{zy} & \varepsilon_{zz} \end{bmatrix}. \tag{3.20}$$

The elements ε_{ii} are the first derivatives of the displacements u_i so that

$$\varepsilon_{xx} = \frac{\partial u_x}{\partial x}, \quad \varepsilon_{yy} = \frac{\partial u_y}{\partial y}, \quad \text{and} \quad \varepsilon_{zz} = \frac{\partial u_z}{\partial z}. \tag{3.21}$$

The shear strain components $2\varepsilon_{ij} = \gamma_{ij}$ are given by [72]

$$\gamma_{xy} = \frac{\partial u_x}{\partial y} + \frac{\partial u_y}{\partial x}, \quad \gamma_{xz} = \frac{\partial u_x}{\partial z} + \frac{\partial u_z}{\partial x}, \quad \ldots \tag{3.22}$$

If an isotropic material is assumed, the strain tensor is symmetric due to

$$\varepsilon_{xy} = \varepsilon_{yx}, \quad \varepsilon_{xz} = \varepsilon_{zx}, \quad \text{and} \quad \varepsilon_{yz} = \varepsilon_{zy}, \tag{3.23}$$

which means that there are only six different values.

Assuming an isotropic material and after constructing $\mathbf{D}$ with the help of (3.16), the stress tensor without residual stress, can be rewritten in the form

$$\begin{pmatrix} \sigma_{xx} \\ \sigma_{yy} \\ \sigma_{zz} \\ \sigma_{xy} \\ \sigma_{yz} \\ \sigma_{zx} \end{pmatrix} = \frac{E(1-\nu)}{(1+\nu)(1-2\nu)} \begin{pmatrix} 1 & \frac{\nu}{1-\nu} & \frac{\nu}{1-\nu} & 0 & 0 & 0 \\ \frac{\nu}{1-\nu} & 1 & \frac{\nu}{1-\nu} & 0 & 0 & 0 \\ \frac{\nu}{1-\nu} & \frac{\nu}{1-\nu} & 1 & 0 & 0 & 0 \\ 0 & 0 & 0 & \frac{1-2\nu}{2(1-\nu)} & 0 & 0 \\ 0 & 0 & 0 & 0 & \frac{1-2\nu}{2(1-\nu)} & 0 \\ 0 & 0 & 0 & 0 & 0 & \frac{1-2\nu}{2(1-\nu)} \end{pmatrix} \begin{pmatrix} \varepsilon_{xx} - \varepsilon_{0,xx} \\ \varepsilon_{yy} - \varepsilon_{0,yy} \\ \varepsilon_{zz} - \varepsilon_{0,zz} \\ \gamma_{xy} - \gamma_{0,xy} \\ \gamma_{yz} - \gamma_{0,yz} \\ \gamma_{zx} - \gamma_{0,zx} \end{pmatrix}. \tag{3.24}$$

ADVANCED OXIDATION MODEL

3.2.5.2 Visco-Elastic Mechanical Model

The material behavior of oxide and nitride are more realistically described with a visco-elastic model [73, 74], especially with a so-called Maxwell element (see Fig. 3.3), which consists of a spring and a dashpot in series. The characteristics of such a Maxwell element is that it takes the stress relaxation and the stress history into account. Also the actual stress is influenced from both, the strain and the strain rate, and, therefore, the stress is a function of time. The Maxwell element can be mathematically formulated with

$$\frac{d\varepsilon}{dt} - \left(\frac{d\sigma}{dt}\frac{1}{G} + \frac{\sigma}{\gamma}\right) = 0, \tag{3.25}$$

where G is the shear modulus and γ is the (shear) viscosity.

Figure 3.3: Maxwell element: a spring and a dashpot in series.

The analytical solution of (3.25) for the temporal stress evolution as a function of the strain velocity is

$$\sigma(t) = \sigma_0 \cdot \exp\left(-\frac{t-t_0}{\tau_r}\right) + \int_{t_0}^{t} G \cdot \exp\left(-\frac{t-t_0}{\tau_r}\right) \frac{d\varepsilon}{dt} \, d\tau_r, \tag{3.26}$$

where σ_0 is the initial stress at time t_0 and τ_r is the Maxwellian relaxation time constant

$$\tau_r = \frac{\gamma}{G}. \tag{3.27}$$

In (3.26) the first term shows that the initial stress relaxes exponentially with time. The evaluation of the integral part leads to

$$\int_{t_0}^{t} G \cdot \exp\left(-\frac{t-t_0}{\tau_r}\right) \frac{d\varepsilon}{dt} \, d\tau_r = \tau_r G\left(1 - \exp\left(-\frac{t-t_0}{\tau_r}\right)\right) \frac{d\varepsilon}{dt}. \tag{3.28}$$

The visco-elastic model is based on the idea that the dilatational components of the stress, which involve the volumetric expansion or compression, and the deviatoric components which only include the shape modification, can be decoupled [75]. For this purpose the

3.2 Mathematical Formulation

material matrix $\mathbf{D}$ from (3.24) can be split in a dilatation and a deviatoric part [76]

$$\mathbf{D} = \mathbf{D}_{dil} + \mathbf{D}_{dev} = \left(H \begin{bmatrix} 1 & 1 & 1 & 0 & 0 & 0 \\ 1 & 1 & 1 & 0 & 0 & 0 \\ 1 & 1 & 1 & 0 & 0 & 0 \\ 0 & 0 & 0 & 0 & 0 & 0 \\ 0 & 0 & 0 & 0 & 0 & 0 \\ 0 & 0 & 0 & 0 & 0 & 0 \end{bmatrix} + G_{\mathit{eff}} \begin{bmatrix} +\frac{4}{3} & -\frac{2}{3} & -\frac{2}{3} & 0 & 0 & 0 \\ -\frac{2}{3} & +\frac{4}{3} & -\frac{2}{3} & 0 & 0 & 0 \\ -\frac{2}{3} & -\frac{2}{3} & +\frac{4}{3} & 0 & 0 & 0 \\ 0 & 0 & 0 & 1 & 0 & 0 \\ 0 & 0 & 0 & 0 & 1 & 0 \\ 0 & 0 & 0 & 0 & 0 & 1 \end{bmatrix} \right). \quad (3.29)$$

Here H is the bulk modulus

$$H = \frac{E}{3(1-2\nu)}, \quad (3.30)$$

and G_{eff} is the so-called effective shear modulus which is in the elastic case the same as the standard shear modulus

$$G_{\mathit{eff}} = G = \frac{E}{2(1+\nu)}. \quad (3.31)$$

In Maxwell's model the dilatation part is assumed purely elastic, while the deviatoric part is modeled by the Maxwell element. In order to find an uncomplicated Maxwell formulation for the deviatoric part in (3.29), it can be assumed in (3.28) that for a short time period ΔT the strain velocity can be kept constant

$$\frac{d\varepsilon}{dt} = \frac{\varepsilon}{\Delta T}, \quad (3.32)$$

so that (3.28) can be expressed in the form

$$\int_{t_0}^{t} G \cdot \exp\left(-\frac{t-t_0}{\tau_r}\right) \frac{d\varepsilon}{dt} \, d\tau_r = \tau_r G \left(1 - \exp\left(-\frac{t-t_0}{\tau_r}\right)\right) \frac{\varepsilon}{\Delta T}. \quad (3.33)$$

So in the visco-elastic case G_{eff} can be written in the form [77, 78]

$$G_{\mathit{eff}} = G \frac{\tau}{\Delta T} \left(1 - \exp\left(-\frac{\Delta T}{\tau}\right)\right). \quad (3.34)$$

This relationship shows that the Maxwell visco-elasticity can be expressed by an effective shear modulus G_{eff} in the deviatoric part of the material matrix $\mathbf{D}$ (3.24). This means that the only difference in the mechanical model between the elastic and visco-elastic case is the different G_{eff} in the material matrix $\mathbf{D}$. So $\mathbf{D}$ depends in the elastic case only on Young's modulus E and the Poisson ratio ν, and in the visco-elastic case additionally on the Maxwellian relaxation time τ.

ADVANCED OXIDATION MODEL

3.2.5.3 Volume Increase and Mechanics

A very important aspect in the oxidation model is, how the volume increase during oxidation can be brought in relation with the mechanical problem. In three dimensions a volume expansion can be formulated with

$$(1 + \varepsilon_{0,xx})(1 + \varepsilon_{0,yy})(1 + \varepsilon_{0,zz}) = V_{rel} + V_{rel}^{add}, \tag{3.35}$$

where V_{rel} is the normalized volume before expansion and thus V_{rel} is always 1.

By assuming that the volume expansion and the strain is small, the strain terms $\varepsilon_{0,ii} \cdot \varepsilon_{0,jj}$ can be neglected (i and j stands for x, y or z), because they are much smaller than the terms $\varepsilon_{0,ii}$. Therefore, with the start volume $V_{rel} = 1$, (3.35) can be reduced to the form

$$\varepsilon_{0,xx} + \varepsilon_{0,yy} + \varepsilon_{0,zz} = V_{rel}^{add}. \tag{3.36}$$

The components $\varepsilon_{0,ii}$ of the residual strain tensor $\tilde{\varepsilon}_0$ are linearly proportional to the normalized additional volume as calculated in (3.8)

$$\varepsilon_{0,ii} = \tfrac{1}{3} V_{rel}^{add}, \tag{3.37}$$

which loads the mechanical problem (3.15) for calculating the displacements and stresses.

3.3 Model Overview

The whole advanced oxidation model is based on a few main equations. The first one describes the oxidant diffusion

$$D(T)\,\Delta C(\vec{x}, t) = k(\eta)\,C(\vec{x}, t), \tag{3.38}$$

and the next equation treats the dynamics of η with

$$\frac{\partial \eta(\vec{x}, t)}{\partial t} = -\frac{1}{\lambda} k(\eta)\,C(\vec{x}, t)/N_1, \tag{3.39}$$

as described in Section 3.2.1 and Section 3.2.2, respectively.

Because of the diffuse interface concept the volume increase of the generating oxide occurs only successively and not abruptly. As explained in Section 3.2.3, the volume increase of the oxidized material is calculated with the η and C values. After a time Δt the normalized additional volume is determined by

$$V_{rel}^{add} = \frac{\lambda - 1}{\lambda} \Delta t\; k(\eta) C(\vec{x}, t)/N_1. \tag{3.40}$$

The normalized additional volume directly loads the mechanical problem

$$\tilde{\sigma} = \mathbf{D}(\tilde{\varepsilon} - \tilde{\varepsilon}_0) + \tilde{\sigma}_0, \tag{3.41}$$

because the principal axis components of the residual strain tensor $\tilde{\varepsilon}_0$ are linearly proportional to V_{rel}^{add} in the form

$$\varepsilon_{0,xx} = \varepsilon_{0,yy} = \varepsilon_{0,zz} = \frac{1}{3} V_{rel}^{add}. \tag{3.42}$$

The introduced so-called effective shear modulus $G_{\textit{eff}}$ in $\mathbf{D}$ (see Section 3.2.5.2) can handle elastic and visco-elastic materials.

Chapter 4

Oxidation of Doped Silicon

THE DOPANT DISTRIBUTION in silicon is stronly influenced by thermal oxidation, because the dopants are redistributed by diffusion and segregation, especially near the silicon wafer surface [79]. However, this dopant redistribution is not the only effect of an oxidation step. Because of the oxide growth, the upper silicon zones are converted into SiO_2 and the Si/SiO_2 interface is moving into deeper silicon zones. Before oxidation, the dopant distribution exhibits generally a Gaussian-like profile, which means that the dopant concentration decreases stronly with the distance from the surface. Therefore, oxidation leads to a general decrease of the dopant concentration at the silicon surface. Furthermore, the formed oxide absorbs the dopants from the converted silicon material. This oxide doping influences the segregation of the dopant concentration at the Si/SiO_2 interface.

An influence of the dopants on the oxide growth rate was only found at very high dopant concentrations near the repective solubility limits of the used doping material, which are in the order of 10^{20} atoms/cm^3 [80]. A high dopant concentration at the silicon surface beneath the SiO_2 (see Fig. 4.1b) causes crystal defects and so the silicon is easier to oxidize. A high number of dopants in the SiO_2 (see Fig. 4.1a) loosens the material and reduces its density, which enables a better oxidant diffusion through the SiO_2 to the interface. In both cases the oxide growth rate is increased.

Since very high dopant concentrations increase the oxide growth rate, theoretically the accelerated oxide growth at heavily doped zones could be used for selective oxidation. But unfortunately in practice this effect is too small to obtain noticeable differences in the oxide thickness. However, the different oxide growth velocities must be taken into account for an etching process. A faster oxide growth leads to a faster material removal by etching.

OXIDATION OF DOPED SILICON

4.1 Dopant Redistribution

The redistribution process depends on the ratio of the solubility of the doping material in silicon and SiO_2. At the Si/SiO_2 interface the dopants are redistributed by segregation until the ratio of their concentration at the interface is the same as the ratio of their solubility in both materials. The ratio of dopant solubility is expressed by the segregation coefficient m which is [80]

$$m = \frac{\text{solubility in silicon}}{\text{solubility in } SiO_2}. \qquad (4.1)$$

As listed in Table 4.1 there are dopant species which solubilize better in SiO_2 than in silicon ($m < 1$) and species which have a reversed behavior ($m > 1$). In case of $m < 1$, as for Boron, the dopant concentration is enhanced at the SiO_2 side, whereas beneath the interface, there is a dopant depletion at the silicon surface (see Fig. 4.1a). For reversed solubility ratios ($m > 1$, like Phosphorus), only few dopant atoms penetrate the interface. In order to obtain the by m determined concentration ratio at the interface, dopant atoms from deeper silicon zones diffuse back to the surface zone. Therefore, the dopant concentration at the silicon surface is enhanced, as illustrated in Fig. 4.1b. In Fig. 4.1 C_c denotes the dopant concentration in the silicon surface zone before oxidation. x is the distance from the silicon surface.

Table 4.1: Segregation coefficients m for important dopant species in silicon [80].

Dopant species	Bor	Phosphor	Antimon	Arsen	Gallium
m	0.1–0.3	10	10	10	20

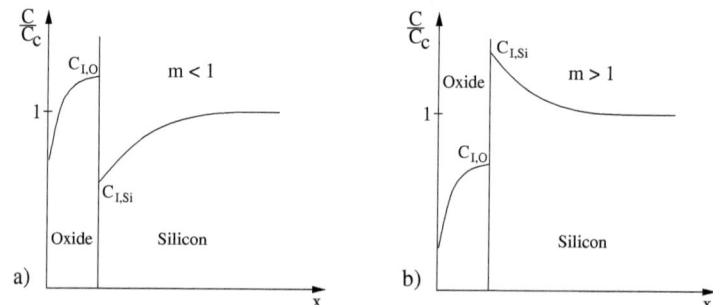

Figure 4.1: Schematic illustration of dopant redistribution.

The dopand redistribution for the moving Si/SiO_2 interface can be described with a diffusion model as presented in the next section.

4.2 Five-Stream Dunham Diffusion Model

Dunham presented 1992 a general model [81] for the coupled diffusion of dopants with point defects, which includes the reaction of dopant-defect pairs with defects and other pairs, as well as all possible charge states for both dopants and pairs. It consists of five streams, because the comprehensive modeling of dopant behavior requires five differential equations, each treating a different concentration stream: one for the dopant atoms, two for the interstitial and vacancy point-defects, and two for the dopant-vacancy and dopant-interstitial pairs [24].

4.2.1 Interaction of Dopants

In silicon a dopant diffuses via interactions with point-defects, which can be described by a set of reactions. First, there are the dopant-defect pairing reactions

$$A^+ + I^i \Longleftrightarrow (AI)^{i+1}, \tag{4.2}$$

$$A^+ + V^i \Longleftrightarrow (AV)^{i+1}, \tag{4.3}$$

where A^+ represent the ionized dopant atoms, I and V represent the interstitials and vacancies, (AI) and (AV) represent the dopant-defect pairs, and i stands for the charge state of the defect or pair as $-$, 0, $+$. Next, the recombination and generation of Frenkel pairs must be considered

$$I^i + V^j \Longleftrightarrow (-i-j)\,e^-, \tag{4.4}$$

where e are electrons. A Frenkel pair is a vacancy-interstitial pair formed when an atom is displaced from a lattice site to an interstitial site.

Additionally, the pairs can interact directly with the opposite type defect to produce a reaction which is equivalent to a pair dissociation followed by defect recombination

$$(AI)^i + V^j \Longleftrightarrow A^+ + (1-i-j)\,e^-, \tag{4.5}$$

$$(AV)^i + I^j \Longleftrightarrow A^+ + (1-i-j)\,e^-, \tag{4.6}$$

Finally, two opposite type pairs can recombine leaving two unpaired dopant atoms

$$(AI)^i + (AV)^j \Longleftrightarrow 2A^+ + (2-i-j)\,e^-. \tag{4.7}$$

The last three reactions provide an alternative path for the recombination and generation of vacancies and interstitials with the potential for a significant increase of the effective recombination rate for Frenkel pairs.

OXIDATION OF DOPED SILICON

4.2.2 Continuity Equations

The five continuity equations for the total concentrations C_X (X stands for A^+, AI, AV, I, or V), over all charge states for a single donor species are [82]

$$\frac{\partial C_{A^+}}{\partial t} = -R_{AI} - R_{AV} + R_{AI+AV} + 2R_{AI+AV}, \tag{4.8}$$

$$\frac{\partial C_I}{\partial t} = -\nabla J_I + R_{AI} - R_{I+V} - R_{AV+I}, \tag{4.9}$$

$$\frac{\partial C_V}{\partial t} = -\nabla J_V + R_{AV} - R_{I+V} - R_{AI+V}, \tag{4.10}$$

$$\frac{\partial C_{AI}}{\partial t} = -\nabla J_{AI} + R_{AI} - R_{AI+V} - R_{AI+AV}, \tag{4.11}$$

$$\frac{\partial C_{AV}}{\partial t} = -\nabla J_{AV} + R_{AV} - R_{AV+I} - R_{AI+AV}. \tag{4.12}$$

R_{AI} and R_{AV} are the net rates of the dopant-defect pairing reactions (4.2) and (4.3) as defined in [81]:

$$R_{AI} = \left[\sum_i k_{AI}^i K_I^i \left(\frac{n_i}{n}\right)^i\right] \left[C_{A^+} C_{I^0} - \frac{C_{(AI)^+}}{K_{A+I}^0}\right], \tag{4.13}$$

$$R_{AV} = \left[\sum_i k_{AV}^i K_V^i \left(\frac{n_i}{n}\right)^i\right] \left[C_{A^+} C_{V^0} - \frac{C_{(AV)^+}}{K_{A+V}^0}\right]. \tag{4.14}$$

k_X are the forward reaction rate coefficients and K_X are the equilibrium constants. n and n_i are the local and intrinsic carrier concentrations.

The net rate R_{I+V} of Frenkel pair recombination (4.4) is [81]

$$R_{AV} = \left[\sum_{i,j} k_{I+V}^{i,j} K_I^i K_V^j \left(\frac{n_i}{n}\right)^{i+j}\right] [C_{I^0} C_{V^0} - C_{I^0}^* C_{V^0}^*], \tag{4.15}$$

where * indicates equilibrium values.

Finally, R_{AI+V}, R_{AV+I}, and R_{AI+AV} are the net rates of the pair-defect (4.5) (4.6) and pair-pair reactions (4.7) [81]:

$$R_{AI+V} = \left[\sum_{i,j} k_{AI+V}^{i,j} K_{AI}^i K_V^j \left(\frac{n_i}{n}\right)^{i+j}\right] \left[C_{(AI)^+} C_{V^0} - K_{A+I}^0 C_{I^0}^* C_{V^0}^* C_{A^+}\right], \tag{4.16}$$

$$R_{AV+I} = \left[\sum_{i,j} k_{AV+I}^{i,j} K_{AV}^i K_I^j \left(\frac{n_i}{n}\right)^{i+j}\right] \left[C_{(AV)^+} C_{I^0} - K_{A+V}^0 C_{I^0}^* C_{V^0}^* C_{A^+}\right], \tag{4.17}$$

$$R_{AI+AV} = \left[\sum_{i,j} k_{AI+AV}^{i,j} K_{AI}^i K_{AV}^j \left(\frac{n_i}{n}\right)^{i+j}\right] \left[C_{(AI)^+} C_{(AV)^+} - K_{A+I}^0 K_{A+V}^0 C_{I^0}^* C_{V^0}^* (C_{A^+})^2\right]. \tag{4.18}$$

The continuity equations (4.30)–(4.30) also need the fluxes of mobile dopants, defects, and pairs. The total flux of interstitials is [81]

$$J_I = -\left[\sum_i D_{I^i} K_I^i \left(\frac{n_i}{n}\right)^i\right] \nabla C_{I^0}, \tag{4.19}$$

where D_{I^i} represents the diffusivity of interstitials of charge state i. Similarly, the total vacancy flux is [81]

$$J_V = -\left[\sum_i D_{V^i} K_V^i \left(\frac{n_i}{n}\right)^i\right] \nabla C_{V^0}. \tag{4.20}$$

The total pair fluxes are [81]

$$J_{AI} = -\left[\sum_i D_{AI^{i+1}} K_{AI}^i \left(\frac{n_i}{n}\right)^i\right]\left[\nabla C_{(AI)^i} + C_{(AI)^i}\left(\frac{n_i}{n}\right)\nabla\left(\frac{n}{n_i}\right)\right], \tag{4.21}$$

$$J_{AV} = -\left[\sum_i D_{AV^{i+1}} K_{AV}^i \left(\frac{n_i}{n}\right)^i\right]\left[\nabla C_{(AV)^i} + C_{(AV)^i}\left(\frac{n_i}{n}\right)\nabla\left(\frac{n}{n_i}\right)\right], \tag{4.22}$$

where D_{AI^i} and D_{AV^i} are the diffusivities of dopant-defect pairs with charge i.

4.3 Segregation Interface Condition

If at the Si/SiO$_2$ interface there is a dopant concentration $C_{I,O}$ and $C_{I,Si}$ on the oxide and silicon side, respectively, as illustrated in Fig. 4.1, the segregation coefficient can be written as [83]

$$m = \frac{C_{I,Si}}{C_{I,O}} \tag{4.23}$$

If it is assumed that $C_{I,O} > C_{I,Si}$ the flux of dopants from the SiO$_2$ segment to the silicon segment through the interface is [83]

$$J_S = k_O C_{I,O} - k_{Si} C_{I,Si} = k_O\left(C_{I,O} - \frac{k_{Si}}{k_O}\right) = h\left(C_{I,O} - \frac{C_{I,Si}}{m}\right), \tag{4.24}$$

where k_O and k_{Si} are the reaction rate coefficients in SiO$_2$ and silicon, respectively. h is the interface transfer coefficien which has units of velocity.

In the steady state the interface flux $J_S = 0$ and (4.24) can be transformed to the relationship

$$\frac{C_{I,O}}{C_{I,Si}} = \frac{k_O}{k_{Si}}. \tag{4.25}$$

4.4 Model Overview with Coupled Dopant Diffusion

If the advanced oxidation model with its equations for the oxidant diffusion

$$D(T)\,\Delta C(\vec{x},t) = k(\eta)\,C(\vec{x},t), \tag{4.26}$$

dynamics of η

$$\frac{\partial \eta(\vec{x},t)}{\partial t} = -\frac{1}{\lambda}k(\eta)\,C(\vec{x},t)/N_1, \tag{4.27}$$

and mechanical problem

$$\tilde{\sigma} = \mathbf{D}(\tilde{\varepsilon} - \tilde{\varepsilon}_0) + \tilde{\sigma}_0, \tag{4.28}$$

is coupled with the five-stream diffusion model for the dopant diffusion, its five continuity equations for the species concentrations

$$\frac{\partial C_{A^+}}{\partial t} = -R_{AI} - R_{AV} + R_{AI+AV} + 2R_{AI+AV}, \tag{4.29}$$

$$\frac{\partial C_I}{\partial t} = -\nabla J_I + R_{AI} - R_{I+V} - R_{AV+I}, \tag{4.30}$$

$$\frac{\partial C_V}{\partial t} = -\nabla J_V + R_{AV} - R_{I+V} - R_{AI+V}, \tag{4.31}$$

$$\frac{\partial C_{AI}}{\partial t} = -\nabla J_{AI} + R_{AI} - R_{AI+V} - R_{AI+AV}, \tag{4.32}$$

$$\frac{\partial C_{AV}}{\partial t} = -\nabla J_{AV} + R_{AV} - R_{AV+I} - R_{AI+AV}. \tag{4.33}$$

must be additionally solved.

Chapter 5

Discretization with the Finite Element Method

PARTIAL DIFFERENTIAL EQUATIONS (PDEs) are widely used to describe and model physical phenomena in different engineering fields and so also in microelectronics' fabrication. Only for simple and geometrically well-defined problems analytical solutions can be found, but for the most problems it is impossible. For these problems, also often with several boundary conditions, the solution of the PDEs can only be found with numerical methods.

The most universal numerical method is based on finite elements. This method has a general mathematical fundament and clear structure. Thereby, it can be relative easily applied for all kinds of PDEs with various boundary conditions in nearly the same way. The finite element method (FEM) has its origin in the mechanics and so it is probably the best method for calculating the displacements during oxidation processes [84]. The finite element formulation works on a large number of discretization elements and also on different kinds of meshes within the domain. Furthermore, it also provides good results for a coarse mesh. It can easily handle complicated geometries, variable material characteristics, and different accuracy demands.

5.1 Basics

The basic aim of the finite element method is to solve a PDE, or a system of coupled PDEs, numerically. Instead of finding the analytic solution of the PDE, which is usually a function of the coordinates, it is tried to determine this function values for discrete coordinates on grid points. For this purpose the continuum is discretized with a number of so-called finite elements which results in a mesh with grid nodes. If the finite elements are appropriately

DISCRETIZATION WITH THE FINITE ELEMENT METHOD

small, the solution of the PDE can be approximated with a simple function, the so-called shape function, in each element, which acts as a contribution to the approximation of the global solution of the PDE. All together a linear or non-linear equation system must be obtained. The real advantage is that such (non-)linear equation systems can be quite easily solved today by computer programs.

5.1.1 Mesh Aspects

For solving a PDE numerically, in the first step the simulation domain Ω is divided into a number of M (as possible geometrically simple) elements E_i (like tetrahedrons or cubes). The quality of the approximated solution depends on the size and so on the number of finite elements in the discretized domain. The elements must be located in such a manner, that there are no empty spaces between them and that they do not overlap:

$$\Omega = \bigcup_{i=1}^{M} E_i \quad \text{and} \quad \forall i \neq j : Int(E_i) \cap Int(E_j) = 0. \tag{5.1}$$

Here $Int(E_i)$ is the set of all points in the element E_i, except those which are located on the surface. Furthermore, surface conformity of neighbor elements is demanded. This means that on each surface of an element inside the domain exact only one neighbor element is bordered.

If P is the set of all grid nodes of the discretized domain Ω, than each grid node $p_k \in P$ has a unique global index $k = 1, \ldots, N$. N is the number of all nodes in the whole mesh. A node p_k can also have several local indices.

5.1.2 Shape Function

From the mathematical point of view the shape function shall interpolate the discrete solution function values between the grid nodes. If a PDE is written in the form

$$\mathcal{D}[u](x, y, z) + g(x, y, z) = 0, \tag{5.2}$$

where $\mathcal{D}$ is a second order differential operator, the desired solution of the PDE is a function $u(x, y, z)$, which is approximated by [85]

$$\tilde{u}(x, y, z) = \sum_{j=1}^{N} u_j N_j(x, y, z) = \{u^T\}\{N\} \tag{5.3}$$

The summation is performed over all grid nodes. Thereby, u_j is the respective value of the function at a grid node with the number j and $N_j(x, y, z)$ is the shape function.

5.1 Basics

The shape function can be chosen quite freely, but it must be appropriate for interpolation. In addition every shape function $N_j(x, y, z)$ must have a value of 1 on the grid node j (with coordinates $\vec{p}_j = \{x_j, y_j, z_j\}^T$) and values of 0 on all other grid nodes:

$$N_j(x_i, y_i, z_i) = \begin{cases} 1 \text{ for i = j} \\ 0 \text{ for i} \neq \text{j} \end{cases}. \tag{5.4}$$

In practical applications linear shape functions or polynomials of low order are used. The numbering of the grid nodes should be carried out in a way, that nodes at Dirichlet-boundaries, where the solution of $u(x, y, z)$ is already known, are ranked behind the others. The nodes where the value u_j must be calculated get the indices $j = 1, \ldots, N_A$, the other N_B nodes at the Dirichlet-boundaries are indexed with $j = N_A + 1, \ldots, N_A + N_B$.

In order to get the solution of the PDE, "only" the unknown coefficients u_j in (5.3) must be obtained. For determining the values u_j there are two available ways, the method of Ritz or the method of weighted residuals [86]. The method of Ritz seeks a stationary point of the variational functional [87]. This variational approach leads to an integral formulation of the PDE. The Ritz method can only be applied, if for the boundary problem an equivalent variational formulation exists. The more universal method is based on the weighted residuals and therefore it is used in this work.

5.1.3 Weighted Residual Method

When the approximated solution $\tilde{u}(x, y, z)$ is reinserted in the PDE (5.2), there exists a residual

$$\mathcal{R} = \mathcal{D}[\tilde{u}(x, y, z)] + g(x, y, z). \tag{5.5}$$

If the residual would disappear ($\mathcal{R} = 0$), the exact solution $\tilde{u}(x, y, z) = u(x, y, z)$ would have been found [88]. Since the function space of the approximation solutions $\tilde{U}$ is a subset of the function space of the exact solutions U, such a solution does not exist for the general case. Therefore, it is tried to fulfill the residual condition not exactly, but with N weighted or averaged linearly independent weight functions (also called test functions) $W_i(x, y, z)$, so that [89]

$$\int_\Omega [W_i(x, y, z)\mathcal{D}[\tilde{u}(x, y, z)] + W_i(x, y, z)g(x, y, z)] \, d\Omega = 0, \quad \text{for} \quad i = 1, 2, 3, \ldots N. \tag{5.6}$$

The weight functions W_i must be chosen in a suitable way, because the quality of the solution depends on them. If the weight functions are identical with the shape functions ($W_i(x, y, z) = N_i(x, y, z)$), this approach is called *Galerkin's method* [85].

DISCRETIZATION WITH THE FINITE ELEMENT METHOD

With the approximated solution (5.3) the weighted residual (5.7) can be rewritten in the form

$$\sum_{j=1}^{N} u_j \int_\Omega W_i(x,y,z) \mathcal{D}[N_j(x,y,z)]\, d\Omega + \int_\Omega W_i(x,y,z) g(x,y,z)\, d\Omega = 0. \qquad (5.7)$$

It is possible to determine the N unknown function values u_j on the grid nodes with this system of N equations.

5.2 Discretization with Tetrahedrons

An often used choice is to discretize the solution domain with tetrahedrons. On the one side a tetrahedron is a relative simple element, especially regarding meshing aspects, on the other side it is an efficient element to discretize structures with non-planar surfaces or complex geometries. As shown in Fig. 5.1, this element is limited by four triangles and has four vertexes which are in any case grid nodes in the mesh.

5.2.1 Shape Functions for a Tetrahedron

For using the weighted residual method the shape functions must be continuous on the transition from one element to its neighbor element. Within the elements they must be at least one-time differentiable. The shape functions will be defined locally on the tetrahedron. It should be noted that the global shape function $N_j(x,y,z)$ is assembled from the local shape functions of the elements which share the same node j.

If it is assumed that the discretization is carried out with linear shape functions, the four vertexes used are the four grid nodes on the element. This means that the shape functions must depend on the x-, y-, and z-coordinate linearly, so that in general form it can be expressed by [90]

$$N_i(x,y,z) = a_i + b_i\, x + c_i\, y + d_i\, z \qquad \text{for} \qquad i = 0,1,2,3. \qquad (5.8)$$

i are the numbers of the local grid node. Since for every grid node a separate shape function is needed, there are four shape functions on a four node element. The coefficients a_i, b_i, c_i, and d_i must be determined in such a manner, that the respective shape function N_i fulfills (5.4). This means, that for example the value of the shape function N_0(x,y,z) must be 1 in node P_0 with its coordinates (x_0, y_0, z_0), and 0 in all other nodes. With this

5.2 Discretization with Tetrahedrons

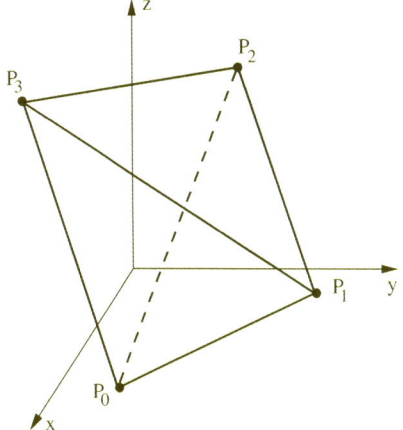

Figure 5.1: Tetrahedral element in a global (x, y, z)-coordinate system.

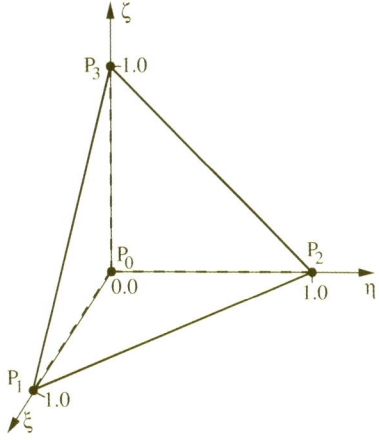

Figure 5.2: Tetrahedral element in a normalized (ξ, η, ζ)-coordinate system.

information the following equation system with N_0 on the four nodes can be written:

$$\begin{aligned}
a_0 + b_0\, x_0 + c_0\, y_0 + d_0\, z_0 &= 1 \\
a_0 + b_0\, x_1 + c_0\, y_1 + d_0\, z_1 &= 0 \\
a_0 + b_0\, x_2 + c_0\, y_2 + d_0\, z_2 &= 0 \\
a_0 + b_0\, x_3 + c_0\, y_3 + d_0\, z_3 &= 0.
\end{aligned} \qquad (5.9)$$

With this equation system it is possible to determine the four unknown coefficients a_0, b_0, c_0, and d_0 for the shape function $N_0(x,y,z)$.

For calculation of the unknown coefficients Cramer's rule can be applied, which says: if there is an equation system $\mathbf{A} \cdot \vec{x} = \vec{b}$, the numbers

$$x_i = \frac{1}{\text{Det}(\mathbf{A})} \begin{vmatrix} A_{11} & \cdots & A_{1\,i-1} & b_1 & A_{1\,i+1} & \cdots & A_{1n} \\ A_{21} & \cdots & A_{2\,i-1} & b_2 & A_{2\,i+1} & \cdots & A_{2n} \\ \vdots & \vdots & \vdots & \vdots & \vdots & \vdots & \vdots \\ A_{n1} & \cdots & A_{n\,i-1} & b_n & A_{n\,i+1} & \cdots & A_{nn} \end{vmatrix} \qquad (5.10)$$

are the components of the solution $\vec{x}$.

With Cramer's rule, for example, the coefficient a_0 from the shape function $N_0(x, y, z)$

DISCRETIZATION WITH THE FINITE ELEMENT METHOD

can be calculated by

$$a_0 = \frac{1}{\text{Det}(\mathbf{D})} \begin{vmatrix} 1 & x_0 & y_0 & z_0 \\ 0 & x_1 & y_1 & z_1 \\ 0 & x_2 & y_2 & z_2 \\ 0 & x_3 & y_3 & z_3 \end{vmatrix}, \qquad (5.11)$$

where the matrix $\mathbf{D}$ is

$$\mathbf{D} = \begin{bmatrix} 1 & x_0 & y_0 & z_0 \\ 1 & x_1 & y_1 & z_1 \\ 1 & x_2 & y_2 & z_2 \\ 1 & x_3 & y_3 & z_3 \end{bmatrix}. \qquad (5.12)$$

By deleting the row i and the column j of a n-rowed determinant, a new (n-1)-rowed sub-determinant α_{ij} with sign (-1^{i+j}) is constructed

$$\alpha_{ij} = (-1)^{i+j} \begin{vmatrix} a_{11} & a_{12} & \cdots & a_{1\,j-1} & a_{1\,j+1} & \cdots & a_{1n} \\ a_{21} & a_{22} & \cdots & a_{2\,j-1} & a_{2\,j+1} & \cdots & a_{2n} \\ \vdots & \vdots & \vdots & \vdots & \vdots & \vdots & \vdots \\ a_{i-1\,1} & a_{i-1\,2} & \cdots & a_{i-1\,j-1} & a_{i-1\,j+1} & \cdots & a_{i-1\,n} \\ a_{i+1\,1} & a_{i+2} & \cdots & a_{i+1\,j-1} & a_{i+1\,j+1} & \cdots & a_{i+1\,n} \\ \vdots & \vdots & \vdots & \vdots & \vdots & \vdots & \vdots \\ a_{n1} & a_{n2} & \cdots & a_{n\,j-1} & a_{n\,j+1} & \cdots & a_{nn} \end{vmatrix}. \qquad (5.13)$$

To simplify the arithmetic, a n-rowed determinant can be calculated with the sum of n (n−1)-rowed sub-determinants (5.13). The determinant can be expanded from a row or column. If for example k is the number of any column, than the determinant is

$$\begin{vmatrix} a_{11} & a_{12} & \cdots & a_{1n} \\ a_{21} & a_{22} & \cdots & a_{2n} \\ \vdots & \vdots & \vdots & \vdots \\ a_{n1} & a_{n2} & \cdots & a_{nn} \end{vmatrix} = \sum_{i=1}^{n} a_{ik}\,\alpha_{ik}. \qquad (5.14)$$

With this rule (5.14) the coefficient a_0 (5.11) can be simplified, because only $a_{0,11} = 1$, and the rest in the first column is 0. After expanding the determinant from the first row one obtains

$$a_0 = \frac{1}{\text{Det}(\mathbf{D})} \begin{vmatrix} x_1 & y_1 & z_1 \\ x_2 & y_2 & z_2 \\ x_3 & y_3 & z_3 \end{vmatrix}. \qquad (5.15)$$

5.2 Discretization with Tetrahedrons

In the same way all other coefficients of the shape function N_0 can be determined

$$b_0 = \frac{-1}{\text{Det}(\mathbf{D})} \begin{vmatrix} 1 & y_1 & z_1 \\ 1 & y_2 & z_2 \\ 1 & y_3 & z_3 \end{vmatrix}, \qquad c_0 = \frac{1}{\text{Det}(\mathbf{D})} \begin{vmatrix} 1 & x_1 & z_1 \\ 1 & x_2 & z_2 \\ 1 & x_3 & z_3 \end{vmatrix}, \qquad d_0 = \frac{-1}{\text{Det}(\mathbf{D})} \begin{vmatrix} 1 & x_1 & y_1 \\ 1 & x_2 & y_2 \\ 1 & x_3 & y_3 \end{vmatrix}. \quad (5.16)$$

The coefficients for the other shape functions can be found in the same way, the only difference leis in the equation system (5.10). Because, for example, $N_1(x, y, z) = a_1 + b_1 x + c_1 y + d_1 z$ must be 1 in node P_1 and 0 in all other nodes, it can be formulated

$$\begin{aligned} a_1 + b_1 x_0 + c_1 y_0 + d_1 z_0 &= 0 \\ a_1 + b_1 x_1 + c_1 y_1 + d_1 z_1 &= 1 \\ a_1 + b_1 x_2 + c_1 y_2 + d_1 z_2 &= 0 \\ a_1 + b_1 x_3 + c_1 y_3 + d_1 z_3 &= 0. \end{aligned} \quad (5.17)$$

With the above described procedure also the coefficients for $N_1(x, y, z)$ can be determined straightforwardly to

$$b_1 = \frac{1}{\text{Det}(\mathbf{D})} \begin{vmatrix} 1 & y_0 & z_0 \\ 1 & y_2 & z_2 \\ 1 & y_3 & z_3 \end{vmatrix}, \qquad \text{and} \qquad c_1 = \frac{-1}{\text{Det}(\mathbf{D})} \begin{vmatrix} 1 & x_0 & z_0 \\ 1 & x_2 & z_2 \\ 1 & x_3 & z_3 \end{vmatrix}. \quad (5.18)$$

5.2.2 Coordinate Transformation

A coordinate transformation can help to simplify the calculation of integrals. For constructing the residual the calculation of the following element integral is frequently needed

$$I_e = \int_T N_i(x, y, z) N_j(x, y, z) \, dz \, dy \, dz \qquad \text{where} \qquad i, j = 0, 1, 2, 3. \quad (5.19)$$

Here the multiplication of two (linear) form functions leads to a more complex polynomial which complicates the integration over the region. It is more practical to integrate over a normalized element T^n (see Fig. 5.2). For this purpose, a tetrahedron with any location in the global (x,y,z)-coordinate system must be transformed into a normalized local (ξ, η, ζ)-coordinate system.

Each point (x,y,z) of the tetrahedral element in the global coordinate system can be transformed to a corresponding point (ξ, η, ζ) in the normalized coordinate system with the following bijective projection rule

$$\begin{aligned} x &= x_0 + (x_1 - x_0)\xi + (x_2 - x_0)\eta + (x_3 - x_0)\zeta, \\ y &= y_0 + (y_1 - y_0)\xi + (y_2 - y_0)\eta + (y_3 - y_0)\zeta, \\ z &= z_0 + (z_1 - z_0)\xi + (z_2 - z_0)\eta + (z_3 - z_0)\zeta. \end{aligned}$$

DISCRETIZATION WITH THE FINITE ELEMENT METHOD

This projection in matrix form leads to

$$\{r\} = \{r_0\} + \mathbf{J} \cdot \{\delta\} \tag{5.20}$$

and the conversion from the global to the normalized coordinates is

$$\{\delta\} = \mathbf{J}^{-1}(\{r\} - \{r_0\}), \tag{5.21}$$

with $\{r\} = (x, y, z)^T$ and $\{\delta\} = (\xi, \eta, \zeta)^T$.
$\mathbf{J}$ is the so-called Jacobian matrix which only depends on the global coordinates (x,y,z)

$$\mathbf{J} = \begin{bmatrix} x_1 - x_0 & x_2 - x_0 & x_3 - x_0 \\ y_1 - y_0 & y_2 - y_0 & y_3 - y_0 \\ z_1 - z_0 & z_2 - z_0 & z_3 - z_0 \end{bmatrix}. \tag{5.22}$$

The element integral (5.19) calculated in the normalized coordinate system must be multiplied with the determinant of the Jacobian matrix

$$I_e = \mathrm{Det}(\mathbf{J}) \int_{T^n} N_i^m(\xi, \eta, \zeta) N_j^m(\xi, \eta, \zeta) \, d\xi \, d\eta \, d\zeta \qquad \text{where} \qquad i, j = 0, 1, 2, 3, \tag{5.23}$$

because the following relationship holds

$$\frac{\partial(\xi, \eta, \zeta)}{\partial(x, y, z)} = \mathrm{Det}(\mathbf{J}). \tag{5.24}$$

The shape functions for the normalized tetrahedron T^n are simpler than those in the global coordinates, because they are reduced to [91]

$$\begin{aligned} N_0^m(\xi, \eta, \zeta) &= 1 - \xi - \eta - \zeta \\ N_1^m(\xi, \eta, \zeta) &= \xi \\ N_2^m(\xi, \eta, \zeta) &= \eta \\ N_3^m(\xi, \eta, \zeta) &= \zeta. \end{aligned} \tag{5.25}$$

These shape functions lead to a simpler integrand in (5.23). A further advantage is that after normalization the lower integration limit is always 0, because as shown in Fig. 5.2, the tetrahedron T^n starts in the origin of ordinates.

Also the upper limits can be found straightforwardly. As shown in Fig. 5.2, the element T^n is bounded by a plane which goes through the points $P_1(1,0,0)$, $P_2(0,1,0)$, and $P_3(0,0,1)$. This plane is described with the equation $\xi + \eta + \zeta = 1$. The maximum on the ξ-axes is 1. The limit in the ξ-η-plane ($\zeta = 0$) can be described with $\eta(\xi) = 1 - \xi$ and the limit in ζ-direction is $\zeta(\xi, \eta) = 1 - \xi - \eta$.

5.2 Discretization with Tetrahedrons

With these limits the element integral of T^n can be written in the form

$$I_e = \text{Det}(\mathbf{J}) \int_{\xi=0}^{1} \int_{\eta=0}^{1-\xi} \int_{\zeta=0}^{1-\xi-\eta} N_i^n \, N_j^n \, d\zeta \, d\eta \, d\xi = \text{Det}(\mathbf{J}) \cdot \begin{cases} \frac{2}{120} & \text{for } i = j \\ \frac{1}{120} & \text{for } i \neq j \end{cases}. \tag{5.26}$$

This is a real advantage of the normalized tetrahedron, because there exists a simple scheme for I_e. The result from integration over the element T^n is either $\frac{1}{60}$ or $\frac{1}{120}$, only depending, if the two functions N_i and N_j are equal or not. So in fact, to calculate the integral I_e in the normalized coordinate system, only the determinant of the Jacobian matrix must be calculated. This procedure is much easier than to find the element integral for a common tetrahedron in the global coordinate system, where each element has a different size. This means that each integral I_e has a different result and must be calculated separately.

5.2.3 Differentiation in the Normalized Coordinate System

The differentiation of the whole projection (5.20), with respect to x leads to

$$\begin{aligned}
1 &= (x_1 - x_0)\frac{\partial \xi}{\partial x} + (x_2 - x_0)\frac{\partial \eta}{\partial x} + (x_3 - x_0)\frac{\partial \zeta}{\partial x}, \\
0 &= (y_1 - y_0)\frac{\partial \xi}{\partial x} + (y_2 - y_0)\frac{\partial \eta}{\partial x} + (y_3 - y_0)\frac{\partial \zeta}{\partial x}, \\
0 &= (z_1 - z_0)\frac{\partial \xi}{\partial x} + (z_2 - z_0)\frac{\partial \eta}{\partial x} + (z_3 - z_0)\frac{\partial \zeta}{\partial x},
\end{aligned} \tag{5.27}$$

with respect to y leads to

$$\begin{aligned}
0 &= (x_1 - x_0)\frac{\partial \xi}{\partial y} + (x_2 - x_0)\frac{\partial \eta}{\partial y} + (x_3 - x_0)\frac{\partial \zeta}{\partial y}, \\
1 &= (y_1 - y_0)\frac{\partial \xi}{\partial y} + (y_2 - y_0)\frac{\partial \eta}{\partial y} + (y_3 - y_0)\frac{\partial \zeta}{\partial y}, \\
0 &= (z_1 - z_0)\frac{\partial \xi}{\partial y} + (z_2 - z_0)\frac{\partial \eta}{\partial y} + (z_3 - z_0)\frac{\partial \zeta}{\partial y},
\end{aligned} \tag{5.28}$$

and with respect to z leads to

$$\begin{aligned}
0 &= (x_1 - x_0)\frac{\partial \xi}{\partial z} + (x_2 - x_0)\frac{\partial \eta}{\partial z} + (x_3 - x_0)\frac{\partial \zeta}{\partial z}, \\
0 &= (y_1 - y_0)\frac{\partial \xi}{\partial z} + (y_2 - y_0)\frac{\partial \eta}{\partial z} + (y_3 - y_0)\frac{\partial \zeta}{\partial z}, \\
1 &= (z_1 - z_0)\frac{\partial \xi}{\partial z} + (z_2 - z_0)\frac{\partial \eta}{\partial z} + (z_3 - z_0)\frac{\partial \zeta}{\partial z}.
\end{aligned} \tag{5.29}$$

DISCRETIZATION WITH THE FINITE ELEMENT METHOD

These derivatives can be also expressed in matrix form

$$\mathbf{I} = \mathbf{J} \times \begin{bmatrix} \frac{\partial \xi}{\partial x} & \frac{\partial \xi}{\partial y} & \frac{\partial \xi}{\partial z} \\ \frac{\partial \eta}{\partial x} & \frac{\partial \eta}{\partial y} & \frac{\partial \eta}{\partial z} \\ \frac{\partial \zeta}{\partial x} & \frac{\partial \zeta}{\partial y} & \frac{\partial \zeta}{\partial z} \end{bmatrix} = \begin{bmatrix} x_1 - x_0 & x_2 - x_0 & x_3 - x_0 \\ y_1 - y_0 & y_2 - y_0 & y_3 - y_0 \\ z_1 - z_0 & z_2 - z_0 & z_3 - z_0 \end{bmatrix} \times \begin{bmatrix} \frac{\partial \xi}{\partial x} & \frac{\partial \xi}{\partial y} & \frac{\partial \xi}{\partial z} \\ \frac{\partial \eta}{\partial x} & \frac{\partial \eta}{\partial y} & \frac{\partial \eta}{\partial z} \\ \frac{\partial \zeta}{\partial x} & \frac{\partial \zeta}{\partial y} & \frac{\partial \zeta}{\partial z} \end{bmatrix}, \quad (5.30)$$

and the following relationship for the partial differential operators in the normalized system can be found with

$$\mathbf{J}^{-1} = \begin{bmatrix} \frac{\partial \xi}{\partial x} & \frac{\partial \xi}{\partial y} & \frac{\partial \xi}{\partial z} \\ \frac{\partial \eta}{\partial x} & \frac{\partial \eta}{\partial y} & \frac{\partial \eta}{\partial z} \\ \frac{\partial \zeta}{\partial x} & \frac{\partial \zeta}{\partial y} & \frac{\partial \zeta}{\partial z} \end{bmatrix}. \quad (5.31)$$

The inverse of the Jacobian matrix (5.22) in the global coordinate system is given by

$$\mathbf{J}^{-1} = \frac{1}{\mathrm{Det}(\mathbf{J})} \begin{bmatrix} L_{11} & L_{12} & L_{13} \\ L_{21} & L_{22} & L_{23} \\ L_{31} & L_{32} & L_{33} \end{bmatrix}, \quad (5.32)$$

where the components L_{ij} of the inverse matrix are

$$\begin{aligned}
L_{11} &= -J_{23} J_{32} + J_{22} J_{33}, \\
L_{12} &= J_{13} J_{32} - J_{12} J_{33}, \\
L_{13} &= -J_{13} J_{22} + J_{12} J_{23}, \\
L_{21} &= J_{23} J_{31} - J_{21} J_{33}, \\
L_{22} &= -J_{13} J_{31} + J_{11} J_{33} \\
L_{23} &= J_{13} J_{21} - J_{11} J_{23}, \\
L_{31} &= -J_{22} J_{31} + J_{21} J_{32} \\
L_{32} &= J_{12} J_{31} - J_{11} J_{23}, \\
L_{33} &= -J_{12} J_{21} + J_{11} J_{22}.
\end{aligned} \quad (5.33)$$

The components J_{ij} of the Jacobian matrix (5.22) depend only on the location of the tetrahedron vertices in the global coordinate system. Because of

$$\frac{\partial \xi}{\partial x} = \frac{L_{11}}{\mathrm{Det}(\mathbf{J})}, \quad \frac{\partial \xi}{\partial y} = \frac{L_{12}}{\mathrm{Det}(\mathbf{J})}, \quad \ldots \quad \frac{\partial \zeta}{\partial z} = \frac{L_{33}}{\mathrm{Det}(\mathbf{J})}, \quad (5.34)$$

there exists a relationship between the partial differential operators in the normalized system and the coordinates of the four nodes from the global system.

5.2 Discretization with Tetrahedrons

If any continuous function $f(\xi, \eta, \zeta)$ in the normalized coordinate system is differentiated in respect to x, then the chain rule must be used so that

$$\frac{\partial f}{\partial x} = \frac{\partial f}{\partial \xi}\frac{\partial \xi}{\partial x} + \frac{\partial f}{\partial \eta}\frac{\partial \eta}{\partial x} + \frac{\partial f}{\partial \zeta}\frac{\partial \zeta}{\partial x}. \tag{5.35}$$

The gradient of the function $f(\xi, \eta, \zeta)$ in the normalized system becomes to

$$\nabla f(\xi, \eta, \zeta) = \begin{bmatrix} \frac{\partial f}{\partial x} \\ \frac{\partial f}{\partial y} \\ \frac{\partial f}{\partial z} \end{bmatrix} = \begin{bmatrix} \frac{\partial f}{\partial \xi}\frac{\partial \xi}{\partial x} + \frac{\partial f}{\partial \eta}\frac{\partial \eta}{\partial x} + \frac{\partial f}{\partial \zeta}\frac{\partial \zeta}{\partial x} \\ \frac{\partial f}{\partial \xi}\frac{\partial \xi}{\partial y} + \frac{\partial f}{\partial \eta}\frac{\partial \eta}{\partial y} + \frac{\partial f}{\partial \zeta}\frac{\partial \zeta}{\partial y} \\ \frac{\partial f}{\partial \xi}\frac{\partial \xi}{\partial z} + \frac{\partial f}{\partial \eta}\frac{\partial \eta}{\partial z} + \frac{\partial f}{\partial \zeta}\frac{\partial \zeta}{\partial z} \end{bmatrix} = (\mathbf{J}^{-1})^{\mathbf{T}} \times \begin{bmatrix} \frac{\partial f}{\partial \xi} \\ \frac{\partial f}{\partial \eta} \\ \frac{\partial f}{\partial \zeta} \end{bmatrix}. \tag{5.36}$$

This means that the gradient operator ∇^n in the normalized system must be multiplied with the transposed of the inverse Jacobian matrix

$$\nabla = (\mathbf{J}^{-1})^{\mathbf{T}} \times \nabla^n = \frac{1}{\text{Det}(\mathbf{J})} \begin{bmatrix} L_{11} & L_{21} & L_{31} \\ L_{12} & L_{22} & L_{32} \\ L_{13} & L_{23} & L_{33} \end{bmatrix} \times \begin{bmatrix} \frac{\partial}{\partial \xi} \\ \frac{\partial}{\partial \eta} \\ \frac{\partial}{\partial \zeta} \end{bmatrix}. \tag{5.37}$$

5.2.4 Discretization of the Oxidant Diffusion

In the continuum formulation from (3.2), the diffusion of oxidants through the oxide material is described with

$$D \Delta C = \eta \, k_{max} \, C. \tag{5.38}$$

When *Galerkin's method* is applied, it is multiplied with a weight function $N_j(x, y, z)$ and integrated over the domain Ω, which leads to

$$\int_\Omega N_j(x,y,z) \, D \, \Delta C \, d\Omega = \int_\Omega N_j(x,y,z) \, \eta \, k_{max} \, C \, d\Omega. \tag{5.39}$$

If there are two functions $u(\vec{x})$ and $v(\vec{x})$ defined on a domain Ω, *Green's theorem* says that

$$\int_\Omega \nabla u \, \nabla v \, d\Omega + \int_\Omega u \, \Delta v \, d\Omega = \int_\Gamma u \frac{\partial v}{\partial n} \, d\Gamma, \tag{5.40}$$

where Γ is the boundary of the domain.

DISCRETIZATION WITH THE FINITE ELEMENT METHOD

With *Green's theorem* the *Galerkin* formulation from (5.39) can be rewritten in the form

$$-D \int_\Omega \nabla N_j \nabla C \, d\Omega + \int_\Gamma N_j \frac{\partial C}{\partial n} \, d\Gamma = k_{max} \int_\Omega N_j \, \eta \, C \, d\Omega. \tag{5.41}$$

Here the diffusion coefficient D and the maximal reaction rate k_{max} do not directly depend on the location and, therefore, they do not need to be integrated over space and can stand outside of the integral. Furthermore, it is assumed that there is no flow of oxidants through the boundary surface and the boundary condition becomes

$$\int_\Gamma N_j \frac{\partial C}{\partial n} \, d\Gamma = 0. \tag{5.42}$$

With this Neumann boundary condition the *Galerkin* formulation for the oxidant diffusion can be reduced to

$$-D \int_\Omega \nabla N_j \nabla C \, d\Omega = k_{max} \int_\Omega N_j \, \eta \, C \, d\Omega \quad \text{for} \quad j = 0, 1, 2, 3. \tag{5.43}$$

With the finite element method it can be assumed that this equation is only valid on a single tetrahedral element T. Furthermore, the scalar function for the oxidant concentration $C(\vec{x}, t)$ is here approximated linearly with

$$C(\vec{x}, t = t_n) = \sum_{i=0}^{3} c_i^{(t_n)} N_i(x, y, z), \tag{5.44}$$

where $c_i^{(t_n)}$ is the value of the oxidant concentration in node i and at discrete time t_n. $N_i(x, y, z)$ is the respective shape function from this node.
The distribution of the normalized silicon $\eta(\vec{x}, t)$ is approximated in the same way so that

$$\eta(\vec{x}, t = t_n) = \sum_{i=0}^{3} \eta_i^{(t_n)} N_i(x, y, z), \tag{5.45}$$

where $\eta_i^{(t_n)}$ is the value of the normalized silicon in node i and at discrete time t_n. $N_i(x, y, z)$ is the linear shape function (5.8) for node i.

With the approximation for $C(\vec{x}, t)$ and $\eta(\vec{x}, t)$, the oxidant diffusion on a single element T can be described with

$$-D \int_T \nabla N_j \nabla \Big(\sum_{i=0}^{3} c_i^{(t_n)} N_i \Big) d\Omega = k_{max} \int_T N_j \sum_{i=0}^{3} \eta_i^{(t_n)} N_i \sum_{i=0}^{3} c_i^{(t_n)} N_i \, d\Omega. \tag{5.46}$$

In the approximation for $C(\vec{x}, t)$ and $\eta(\vec{x}, t)$ the shape function $N_i(x, y, z)$ is the same. Since the values c_i and η_i in the nodes do not depend on the spatial location, (5.46) can be rewritten as

$$-D \sum_{i=0}^{3} \Big(c_i^{(t_n)} \int_T \nabla N_j \nabla N_i \, d\Omega \Big) = k_{max} \sum_{i=0}^{3} \Big(c_i^{(t_n)} \eta_i^{(t_n)} \int_T N_j \, N_i \, d\Omega \Big), \tag{5.47}$$

5.2 Discretization with Tetrahedrons

where it is assumed that $\sum \eta_i N_i \sum c_i N_i \approx \sum \eta_i c_i N_i$.
By substituting the integrals with

$$K_{ij} = \int_T \nabla N_i \, \nabla N_j \, d\Omega$$

$$M_{ij} = \int_T N_i \, N_j \, d\Omega \qquad (5.48)$$

the discretized equation for the oxidant diffusion is simplified to

$$-D \sum_{i=0}^{3} K_{ij} \, c_i^{(t_n)} = k_{max} \sum_{i=0}^{3} M_{ij} \, c_i^{(t_n)} \, \eta_i^{(t_n)} \qquad \text{for} \qquad j = 0, 1, 2, 3. \qquad (5.49)$$

The components M_{ij} were already calculated in Section 5.2.2. After integration of $\int_T N_i N_j \, d\Omega$ in the normalized coordinate system, it was found that

$$M_{ij} = \text{Det}(\mathbf{J}) \cdot \begin{cases} \frac{2}{120} & \text{for } i = j \\ \frac{1}{120} & \text{for } i \neq j \end{cases}. \qquad (5.50)$$

For calculating the components K_{ij} the integral is also transformed from the global to the normalized coordinate system, and with (5.24) follows

$$K_{ij} = \text{Det}(\mathbf{J}) \int_{\xi=0}^{1} \int_{\eta=0}^{1-\xi} \int_{\zeta=0}^{1-\xi-\eta} \nabla N_i^n \, \nabla N_j^n \, d\zeta \, d\eta \, d\xi. \qquad (5.51)$$

where $N_{i,j}^n(\xi, \eta, \zeta)$ are the shape functions for a normalized tetrahedron T^n from (5.26). It was demonstrated in (5.37) that the transformation of the gradient operator ∇ is carried out by a multiplication with the matrix $(\mathbf{J}^{-1})^\mathbf{T}$, so that the integrand from (5.51) becomes

$$\nabla N_i^n \, \nabla N_j^n = (\mathbf{J}^{-1})^\mathbf{T} \, \nabla^n N_i^n \, (\mathbf{J}^{-1})^\mathbf{T} \, \nabla^n N_j^n =$$

$$= \frac{1}{\text{Det}(\mathbf{J})} \begin{bmatrix} L_{11} & L_{21} & L_{31} \\ L_{12} & L_{22} & L_{32} \\ L_{13} & L_{23} & L_{33} \end{bmatrix} \times \begin{bmatrix} \frac{\partial N_i^n}{\partial \xi} \\ \frac{\partial N_i^n}{\partial \eta} \\ \frac{\partial N_i^n}{\partial \zeta} \end{bmatrix} \cdot \frac{1}{\text{Det}(\mathbf{J})} \begin{bmatrix} L_{11} & L_{21} & L_{31} \\ L_{12} & L_{22} & L_{32} \\ L_{13} & L_{23} & L_{33} \end{bmatrix} \times \begin{bmatrix} \frac{\partial N_j^n}{\partial \xi} \\ \frac{\partial N_j^n}{\partial \eta} \\ \frac{\partial N_j^n}{\partial \zeta} \end{bmatrix} = \qquad (5.52)$$

$$= \frac{1}{\text{Det}(\mathbf{J})} \begin{bmatrix} L_{11} \frac{\partial N_i^n}{\partial \xi} + L_{21} \frac{\partial N_i^n}{\partial \eta} + L_{31} \frac{\partial N_i^n}{\partial \zeta} \\ L_{12} \frac{\partial N_i^n}{\partial \xi} + L_{22} \frac{\partial N_i^n}{\partial \eta} + L_{32} \frac{\partial N_i^n}{\partial \zeta} \\ L_{13} \frac{\partial N_i^n}{\partial \xi} + L_{23} \frac{\partial N_i^n}{\partial \eta} + L_{33} \frac{\partial N_i^n}{\partial \zeta} \end{bmatrix} \cdot \frac{1}{\text{Det}(\mathbf{J})} \begin{bmatrix} L_{11} \frac{\partial N_j^n}{\partial \xi} + L_{21} \frac{\partial N_j^n}{\partial \eta} + L_{31} \frac{\partial N_j^n}{\partial \zeta} \\ L_{12} \frac{\partial N_j^n}{\partial \xi} + L_{22} \frac{\partial N_j^n}{\partial \eta} + L_{32} \frac{\partial N_j^n}{\partial \zeta} \\ L_{13} \frac{\partial N_j^n}{\partial \xi} + L_{23} \frac{\partial N_j^n}{\partial \eta} + L_{33} \frac{\partial N_j^n}{\partial \zeta} \end{bmatrix}.$$

DISCRETIZATION WITH THE FINITE ELEMENT METHOD

After the multiplication of the two vectors and rearranging of this scalar product, (5.52) becomes

$$\frac{\partial N_i^n}{\partial \xi}\frac{\partial N_j^n}{\partial \xi}(L_{11}^2 + L_{12}^2 + L_{13}^2) + \frac{\partial N_i^n}{\partial \xi}\frac{\partial N_j^n}{\partial \eta}(L_{11}L_{21} + L_{12}L_{22} + L_{13}L_{23}) +$$

$$+\frac{\partial N_i^n}{\partial \xi}\frac{\partial N_j^n}{\partial \zeta}(L_{11}L_{31} + L_{12}L_{32} + L_{13}L_{33}) + \frac{\partial N_i^n}{\partial \eta}\frac{\partial N_j^n}{\partial \xi}(L_{21}L_{11} + L_{22}L_{12} + L_{23}L_{13}) +$$

$$+\frac{\partial N_i^n}{\partial \eta}\frac{\partial N_j^n}{\partial \eta}(L_{21}^2 + L_{22}^2 + L_{23}^2) + \frac{\partial N_i^n}{\partial \eta}\frac{\partial N_j^n}{\partial \zeta}(L_{21}L_{31} + L_{22}L_{32} + L_{23}L_{33}) + \quad (5.53)$$

$$+\frac{\partial N_i^n}{\partial \zeta}\frac{\partial N_j^n}{\partial \xi}(L_{31}L_{11} + L_{32}L_{12} + L_{33}L_{13}) + \frac{\partial N_i^n}{\partial \zeta}\frac{\partial N_j^n}{\partial \eta}(L_{31}L_{21} + L_{32}L_{22} + L_{33}L_{23}) +$$

$$+\frac{\partial N_i^n}{\partial \zeta}\frac{\partial N_j^n}{\partial \zeta}(L_{31}^2 + L_{32}^2 + L_{33}^2).$$

Here the L_{xy}-terms which only depend on the location of the nodes in the global coordinate system, can be replaced by six constant coefficients $G_A - G_F$

$$\begin{aligned} G_A &= L_{11}^2 + L_{12}^2 + L_{13}^2 \\ G_B &= L_{11}L_{21} + L_{12}L_{22} + L_{13}L_{23} \\ G_C &= L_{11}L_{31} + L_{12}L_{32} + L_{13}L_{33} \\ G_D &= L_{21}^2 + L_{22}^2 + L_{23}^2 \\ G_E &= L_{21}L_{31} + L_{22}L_{32} + L_{23}L_{33} \\ G_F &= L_{31}^2 + L_{32}^2 + L_{33}^2. \end{aligned} \quad (5.54)$$

With (5.54) the scalar product (5.53) can be simplified to

$$\frac{\partial N_i^n}{\partial \xi}\frac{\partial N_j^n}{\partial \xi} G_A + \left(\frac{\partial N_i^n}{\partial \xi}\frac{\partial N_j^n}{\partial \eta} + \frac{\partial N_i^n}{\partial \eta}\frac{\partial N_j^n}{\partial \xi}\right) G_B + \left(\frac{\partial N_i^n}{\partial \xi}\frac{\partial N_j^n}{\partial \zeta} + \frac{\partial N_i^n}{\partial \zeta}\frac{\partial N_j^n}{\partial \xi}\right) G_C +$$

$$+\frac{\partial N_i^n}{\partial \eta}\frac{\partial N_j^n}{\partial \eta} G_D + \left(\frac{\partial N_i^n}{\partial \eta}\frac{\partial N_j^n}{\partial \zeta} + \frac{\partial N_i^n}{\partial \zeta}\frac{\partial N_j^n}{\partial \eta}\right) G_E + \frac{\partial N_i^n}{\partial \zeta}\frac{\partial N_j^n}{\partial \zeta} G_F. \quad (5.55)$$

For example, the simplified scalar product (5.55) for $i = j = 0$, ($N_0 = 1 - \xi - \eta - \zeta$) is

$$G_A + 2G_B + 2G_C + G_D + 2G_E + G_F, \quad (5.56)$$

and for $i = 0, j = 1$ ($N_1 = \xi$) it is

$$-G_A - G_B - G_C. \quad (5.57)$$

For all combinations of $i, j = 0, 1, 2, 3$ the simplified scalar product can be written in the form

$$\mathbf{S_A}\, G_A + \mathbf{S_B}\, G_B + \mathbf{S_C}\, G_C + \mathbf{S_D}\, G_D + \mathbf{S_E}\, G_E + \mathbf{S_F}\, G_F, \quad (5.58)$$

5.2 Discretization with Tetrahedrons

Instead of finding the scalar product (5.55), and the components of the matrices $\mathbf{S_A} - \mathbf{S_F}$ for all combinations of $i, j = 0, 1, 2, 3$ by the way like in (5.56), it is more comfortable to use

$$\mathbf{N^n} = \begin{bmatrix} 1-\xi-\eta-\zeta \\ \xi \\ \eta \\ \zeta \end{bmatrix} \quad \frac{\partial \mathbf{N^n}}{\partial \xi} = \begin{bmatrix} -1 \\ 1 \\ 0 \\ 0 \end{bmatrix} \quad \frac{\partial \mathbf{N^n}}{\partial \eta} = \begin{bmatrix} -1 \\ 0 \\ 1 \\ 0 \end{bmatrix} \quad \frac{\partial \mathbf{N^n}}{\partial \zeta} = \begin{bmatrix} -1 \\ 0 \\ 0 \\ 1 \end{bmatrix}, \quad (5.59)$$

and get the matrices $\mathbf{S_A} - \mathbf{S_F}$ with

$$\mathbf{S_A} = \frac{\partial \mathbf{N^n}}{\partial \xi} \times \left(\frac{\partial \mathbf{N^n}}{\partial \xi}\right)^{\mathbf{T}} = \begin{bmatrix} -1 \\ 1 \\ 0 \\ 0 \end{bmatrix} \times \begin{bmatrix} -1 \\ 1 \\ 0 \\ 0 \end{bmatrix}^{\mathbf{T}} = \begin{bmatrix} 1 & -1 & 0 & 0 \\ -1 & 1 & 0 & 0 \\ 0 & 0 & 0 & 0 \\ 0 & 0 & 0 & 0 \end{bmatrix}, \quad (5.60)$$

$$\mathbf{S_B} = \begin{bmatrix} -1 \\ 1 \\ 0 \\ 0 \end{bmatrix} \times \begin{bmatrix} -1 \\ 0 \\ 1 \\ 0 \end{bmatrix}^{\mathbf{T}} + \begin{bmatrix} -1 \\ 0 \\ 1 \\ 0 \end{bmatrix} \times \begin{bmatrix} -1 \\ 1 \\ 0 \\ 0 \end{bmatrix}^{\mathbf{T}} = \begin{bmatrix} 2 & -1 & -1 & 0 \\ -1 & 0 & 1 & 0 \\ -1 & 1 & 0 & 0 \\ 0 & 0 & 0 & 0 \end{bmatrix}, \ldots \quad (5.61)$$

Because the derivatives of the linear shape functions $N_{i,j}^n(\xi, \eta, \zeta)$ can only result in the values -1, 0 or $+1$, the integral from (5.51) becomes

$$\int_{\xi=0}^{1} \int_{\eta=0}^{1-\xi} \int_{\zeta=0}^{1-\xi-\eta} \frac{\partial N_i^n}{\partial \alpha} \frac{\partial N_j^n}{\partial \beta} d\zeta \, d\eta \, d\xi = \int_{\xi=0}^{1} \int_{\eta=0}^{1-\xi} \int_{\zeta=0}^{1-\xi-\eta} \pm 1 \, d\zeta \, d\eta \, d\xi = \pm \frac{1}{6}, \quad (5.62)$$

which means that all matrices $\mathbf{S_A} - \mathbf{S_F}$ must be weighted with $\frac{1}{6}$.

After finding $\nabla N_i^n \nabla N_j^n$ and integration over the (normalized) element, the coefficients K_{ij} are

$$K_{ij} = \frac{1}{6}(\mathbf{S_{A},}_{ij} G_A + \mathbf{S_{B,}}_{ij} G_B + \mathbf{S_{C,}}_{ij} G_C + \mathbf{S_{D,}}_{ij} G_D + \mathbf{S_{E,}}_{ij} G_E + \mathbf{S_{F,}}_{ij} G_F)/\text{Det}(\mathbf{J}). \quad (5.63)$$

Galerkin's method assumes that the residual from the (discretized) equation of the oxidant diffusion is zero, and so (5.49) is rewritten in the form

$$\sum_{i=1}^{4} \left(D K_{ij} c_i^{(t_n)} + k_{max} M_{ij} c_i^{(t_n)} \eta_i^{(t_n)} \right) = 0, \qquad j = 0, 1, 2, 3. \quad (5.64)$$

This is a system with four equations, but with eight unknown variables $c_i^{(t_n)}$ and $\eta_i^{(t_n)}$. Because of more unknowns than equations it is impossible to solve this equation system in present form. In the next section the required four equations are introduced.

DISCRETIZATION WITH THE FINITE ELEMENT METHOD

5.2.5 Discretization of the η-Dynamics

The dynamics of the normalized silicon concentration η in the continuum formulation (3.5) is described by

$$\frac{\partial \eta}{\partial t} = -\frac{1}{\lambda}\eta\, k_{max}\, C/N_1, \tag{5.65}$$

After applying *Galerkin's method* with a weight function $N_j(x,y,z)$ one obtains in a domain Ω is

$$\int_\Omega N_j \frac{\partial \eta}{\partial t}\, d\Omega = -K_A \int_\Omega N_j\, \eta\, C\, d\Omega \quad \text{with} \quad K_A = \frac{k_{max}}{\lambda N_1} \quad \text{and} \quad j=0,1,2,3. \tag{5.66}$$

Because of the time dependence in this equation an additional time discretization of the term $\frac{\partial \eta(\vec{x},t)}{\partial t}$ is necessary. This time discretization is performed with a simple backward-Euler method as

$$\frac{\partial \eta(\vec{x}, t=t_n)}{\partial t} = \frac{\eta(\vec{x},t_n) - \eta(\vec{x},t_{n-1})}{\Delta t}, \tag{5.67}$$

where t_n and t_{n-1} are two successive discrete times.

For an equation $\frac{\partial y(t)}{\partial t} = f(y,t)$ two Euler methods can be applied. The forward-Euler method is an explicit simple method, because the new value $y^{(t_{n+1})} = y^{t_n} + \Delta t \cdot f(t_n, y^{t_n})$ is defined in terms of values that are already known [92]. The backward-Euler method comes from using $f(y,t)$ at the end of a time step, when $t=t_{n+1}$. It is an implicit method, because in order to obtain the new discrete value $y^{(t_{n+1})}$ a linear equation of the form $y^{(t_{n+1})} = y^{t_n} + \Delta t \cdot f(t_{n+1}, y^{(t_{n+1})})$ must be solved [92], which requires additional computing time. But compared with the forward-Euler method the most important advantage of the backward-Euler method is that a much larger time step size Δt can be used. The reason is that implicit methods are usually much more stable for solving a stiff equation. A stiff equation is a differential equation for which certain numerical methods for solving the equation are numerically unstable, unless the time step size is taken to be extremely small [93].

The spatial approximation for the oxdiant concentration $C(\vec{x},t)$ and the normalized silicon $\eta(\vec{x},t)$ for one finite element T is the same as in Section 5.2.4

$$C(\vec{x}, t=t_n) = \sum_{i=0}^{3} c_i^{(t_n)}\, N_i(x,y,z), \tag{5.68}$$

$$\eta(\vec{x}, t=t_n) = \sum_{i=0}^{3} \eta_i^{(t_n)}\, N_i(x,y,z). \tag{5.69}$$

5.2 Discretization with Tetrahedrons

With the time discretization (5.67) and the spatial approximation for $C(\vec{x},t)$ and $\eta(\vec{x},t)$, the *Galerkin* formulation for the η-dynamics on a finite element T becomes

$$\int_T N_j \frac{1}{\Delta t}\Big(\sum_{i=0}^{3}\eta_i^{(t_n)}N_i - \sum_{i=0}^{3}\eta_i^{(t_{n-1})}N_i\Big)\,d\Omega = -K_A \int_T N_j \sum_{i=0}^{3}\eta_i^{(t_n)}N_i \sum_{i=0}^{3}c_i^{(t_n)}N_i\,d\Omega, \quad (5.70)$$

Because of the same shape function N_i for $C(\vec{x},t)$ and $\eta(\vec{x},t)$ and no spatial dependence of $c_i^{(t_n)}$ and $\eta_i^{(t_n)}$, the last equation can be rearranged to

$$\frac{1}{\Delta t}\sum_{i=0}^{3}\Big((\eta_i^{(t_n)}-\eta_i^{(t_{n-1})})\int_T N_i N_j\,d\Omega\Big) = -K_A \sum_{i=0}^{3}\Big(\eta_i^{(t_n)}c_i^{(t_n)}\int_T N_i N_j\,d\Omega\Big). \quad (5.71)$$

After substituting $M_{ij} = \int_T N_i N_j\,d\Omega$ from (5.50) the discretized equation for the η-dynamics is simplified to

$$\frac{1}{\Delta t}\sum_{i=0}^{3}M_{ij}\big((\eta_i^{(t_n)}-\eta_i^{(t_{n-1})})\big) = -K_A \sum_{i=0}^{3}M_{ij}\eta_i^{(t_n)}c_i^{(t_n)}. \quad (5.72)$$

In order to fulfill *Galerkin's* demand that the residual should be zero, the last equation can be rewritten as

$$\sum_{i=0}^{3}\Big(M_{ij}\big(\eta_i^{(t_n)}-\eta_i^{(t_{n-1})}\big) + K_A M_{ij}\eta_i^{(t_n)}c_i^{(t_n)}\Delta t\Big) = 0 \quad \text{for} \quad j=0,1,2,3, \quad (5.73)$$

which is also a system with four equations and eight unknown variables $c_i^{(t_n)}$ and $\eta_i^{(t_n)}$. The values for $\eta_i^{(t_{n-1})}$ are already determined at the previous time step.

By combining the two equation systems (5.64) and (5.73), a non-linear but fully determined equation system for one finite element, with 8 equations and the 8 unknowns $c_0^{(t_n)} - c_3^{(t_n)}$ and $\eta_0^{(t_n)} - \eta_3^{(t_n)}$, is obtained. The system is non-linear because of the product $\eta_i^{(t_n)}c_i^{(t_n)}$ in (5.64) and in (5.73). The complete equation system can be solved (for example with the Newton method) at each time point t_n and the values for $c_i^{(t_n)}$ and $\eta_i^{(t_n)}$ can be determined.

5.2.6 Discretization of the Mechanics

The main interest in the continuum mechanics is the deformation of a body by internal or external forces. The deformation is expressed by the displacements $d(x,y,z)$. The displacement of a point in a three-dimensional elastic continuum is defined by three displacement components $u(x,y,z)$, $u(x,y,z)$, and $u(x,y,z)$ in directions of the three coordinates x, y, and z, so that

$$\vec{d}(x,y,z) = \left\{\begin{array}{c} u(x,y,z) \\ v(x,y,z) \\ w(x,y,z) \end{array}\right\}. \quad (5.74)$$

DISCRETIZATION WITH THE FINITE ELEMENT METHOD

In contrast to the previous differential equations (5.38) and (5.65) for the mechanics *Galerkin's method* is not needed. Instead the virtual work concept is used [94]. The displacement components u, v, and w are directly discretized on a finite tetrahedral element

$$u(x,y,z) = \sum_{i=0}^{3} u_i \, N_i(x,y,z),$$

$$v(x,y,z) = \sum_{i=0}^{3} v_i \, N_i(x,y,z), \qquad (5.75)$$

$$w(x,y,z) = \sum_{i=0}^{3} w_i \, N_i(x,y,z),$$

where u_i, v_i, and w_i are the displacement values in x-, y-, and z-direction on node i and N_i is the linear shape function from (5.8).

The strain components in the elastic case are first order derivatives of the displacement components

$$\tilde{\varepsilon}^e = \begin{Bmatrix} \varepsilon_{xx} \\ \varepsilon_{yy} \\ \varepsilon_{zz} \\ \gamma_{xy} \\ \gamma_{yz} \\ \gamma_{zx} \end{Bmatrix} = \begin{Bmatrix} \dfrac{\partial u}{\partial x} \\ \dfrac{\partial v}{\partial y} \\ \dfrac{\partial w}{\partial z} \\ \dfrac{\partial u}{\partial y} + \dfrac{\partial v}{\partial x} \\ \dfrac{\partial v}{\partial z} + \dfrac{\partial w}{\partial y} \\ \dfrac{\partial w}{\partial x} + \dfrac{\partial u}{\partial z} \end{Bmatrix} = \mathbf{B}\,\vec{d^e} = [\mathbf{B_0, B_1, B_2, B_3}]\,\vec{d^e}. \qquad (5.76)$$

The element displacement is defined by the twelve displacement components of the four nodes as

$$\vec{d^e} = \begin{Bmatrix} \vec{d_0} \\ \vec{d_1} \\ \vec{d_2} \\ \vec{d_3} \end{Bmatrix} \quad \text{with} \quad \vec{d_0} = \begin{Bmatrix} u_0 \\ v_0 \\ w_0 \end{Bmatrix} \quad \text{etc.} \qquad (5.77)$$

5.2 Discretization with Tetrahedrons

The submatrix $\mathbf{B_i}$ of displacement derivatives for the node i is [95]

$$\mathbf{B_i} = \begin{bmatrix} \frac{\partial N_i}{\partial x} & 0 & 0 \\ 0 & \frac{\partial N_i}{\partial y} & 0 \\ 0 & 0 & \frac{\partial N_i}{\partial z} \\ \frac{\partial N_i}{\partial y} & \frac{\partial N_i}{\partial x} & 0 \\ 0 & \frac{\partial N_i}{\partial z} & \frac{\partial N_i}{\partial y} \\ \frac{\partial N_i}{\partial z} & 0 & \frac{\partial N_i}{\partial x} \end{bmatrix} = \begin{bmatrix} b_i & 0 & 0 \\ 0 & c_i & 0 \\ 0 & 0 & d_i \\ c_i & b_i & 0 \\ 0 & d_i & c_i \\ d_i & 0 & b_i \end{bmatrix}, \quad \text{where} \quad i = 0, 1, 2, 3. \tag{5.78}$$

The coefficients b_i, c_i, and d_i are the same as already presented in (5.16) and (5.18).

The entire inner virtual work on a continuous elastic body, and so also on a finite element can be written in the form [94]

$$W_{inner} = \int_V \{\tilde{\varepsilon}^e\}^T \sigma^e dV, \tag{5.79}$$

with the stress tensor $\tilde{\sigma}$ (3.15). Here it is assumed that there is no residual stress $\tilde{\sigma}_0$, and the stress tensor is

$$\tilde{\sigma} = \mathbf{D}(\tilde{\varepsilon} - \tilde{\varepsilon}_0). \tag{5.80}$$

In discretized form the transposed strain tensor is

$$\{\tilde{\varepsilon}^e\}^T = \vec{d}^{e^T} \mathbf{B^T}. \tag{5.81}$$

After discretization the stress tensor (5.80) can be arranged as a function of the element displacement vector

$$\sigma^e = \mathbf{D}(\tilde{\varepsilon}^e - \tilde{\varepsilon}_0^e) = \mathbf{D}\,\mathbf{B}\,\vec{d}^e - \mathbf{D}\,\tilde{\varepsilon}_0^e. \tag{5.82}$$

Together with the transposed strain tensor (5.82), this stress tensor leads to the following discretized form of the equation for the inner virtual work on a finite element

$$W_{inner} = \vec{d}^{e^T} \int_V \left(\mathbf{B^T D B}\,\vec{d}^e - \mathbf{B^T D}\,\tilde{\varepsilon}_0^e \right) dV. \tag{5.83}$$

The outer virtual work on a finite element, caused by the external nodal forces is

$$W_{outer} = \vec{d}^{e^T} \vec{f}^{e^{ext}} = 0, \tag{5.84}$$

DISCRETIZATION WITH THE FINITE ELEMENT METHOD

because it is assumed that during the oxidation process there are not external forces acting.

On any elastic body, and so also on a finite element, the inner work must be equal with the outer work

$$W_{inner} = \vec{d^e}^T \int_V \left(\mathbf{B^T D B} \, \vec{d^e} - \mathbf{B^T D} \, \tilde{\varepsilon}_0^{\,e} \right) dV = 0 = W_{outer}, \tag{5.85}$$

which can be simplified to

$$\int_V \mathbf{B^T D B} \, \vec{d^e} \, dV = \int_V \mathbf{B^T D} \, \tilde{\varepsilon}_0^{\,e} \, dV. \tag{5.86}$$

Here the integrals can be substituted as sketched in [95]

$$\mathbf{K^e} = \int_V \mathbf{B^T D B} \, dV = \mathbf{B^T D B} \, V^e, \tag{5.87}$$

$$\vec{f^e_{int}} = \int_V \mathbf{B^T D} \, \tilde{\varepsilon}_0^{\,e} \, dV = \mathbf{B^T D} \, \tilde{\varepsilon}_0^{\,e} \, V^e, \tag{5.88}$$

where $\mathbf{K^e}$ is the so-called stiffness matrix and $\vec{f^e_{int}}$ can be declared as internal force vector. Since the integrands are not functions of the x-, y-, or z-coordinates, the integration over the volume is equal with its much more simpler multiplication. The volume of any tetrahedron in the global coordinate system can be calculated with the determinant of matrix (5.12) by

$$V^e = \frac{1}{6} \mathrm{Det}(\mathbf{D}). \tag{5.89}$$

The most important fact is that the residual strain tensor $\tilde{\varepsilon}_0^{\,e}$ in (5.88) loads the mechanical system. Because the residual strain components $\varepsilon_{0,ii}$ are directly proportional to the normalized additional volume (3.37), there is a relationship between the volume expansion and the internal nodal forces.

With the integral substitutions (5.87) and (5.88), the balance equation (5.86) becomes a linear equation system for the mechanical problem on one finite element

$$\mathbf{K^e} \, \vec{d^e} = \vec{f^e}. \tag{5.90}$$

The system is fully determined, because there are 12 equations and also 12 unknown displacement-components (three on each node) on the tetrahedron.

5.3 Assembling and Solving

Regarding assembling the finite element method is based on the principle that the components of the local element matrices must be assembled to a global matrix for building up a global equation system. Only with the global system all unknown variables on the grid nodes in the discretized domain can be determined. Each global grid node is shared by a varible number of finite elements which all make a contribution to the solution of the unknown values on the involved nodes.

5.3.1 Principle of Assembling

The assembling procedure from a local element matrix $\mathbf{A}^e$ to a global matrix $\mathbf{A}^g$ has the same routines for two- and three-dimensional structures. Therefore, the assembling procedure is demonstrated on a simpler two-dimensional example as shown in Fig. 5.3. The dimension of the local matrix $\mathbf{A}^e$ is always $nk \times nk$, where n is the number of grid nodes on the finite element ($n = 3$ for triangles and $n = 4$ for tetrahedrons) and k is the number of unknown variables on a grid node. The dimension of the global matrix $\mathbf{A}^g$ is always $Nk \times Nk$, where N is the total number of grid nodes in the discretized domain. If it is assumed that there is only one sought variable φ

$$\mathbf{A}^e \varphi^e = b^e, \qquad \mathbf{A}^g \varphi^g = b^g \tag{5.91}$$

the dimension of $\mathbf{A}^e$ is 3×3 in the two-dimensional case and the dimension of $\mathbf{A}^g$ is $N \times N$.

The local matrix uses the local node indexes which are 1, 2, and 3 for every finite element. The global indexes for these grid nodes are different. There must exists a transformation $T(\mathbf{A}^e)$ which projects the local indexes k and l of the components A^e_{kl} to the global indexes i and j. For example, the element 66 in Fig. 5.3 with its local nodes 1, 2, and 3 has the global nodes 42, 30, and 25 and the index transformation is

$$1 \to 42, \qquad 2 \to 30, \qquad \text{and} \qquad 3 \to 25. \tag{5.92}$$

This means that the components of the local matrix $\mathbf{A}^{66}$ from element 66 are transformed to the global matrix $\mathbf{A}^g$ in the way

$$A^{66}_{1,1} \to A^g_{42,42}, \qquad A^{66}_{1,2} \to A^g_{42,30}, \qquad A^{66}_{1,3} \to A^g_{42,25}, \qquad A^{66}_{2,1} \to A^g_{30,42}, \qquad \ldots \tag{5.93}$$

Another important aspect is that a global grid node is shared by a number of different finite elements. This fact is taken into account during assembling of the global matrix by the so-called superposition principle. This means that the components A^g_{ij} of the global matrix are summed up from the contributions A^e_{kl} of the local element matrices. For example, the global grid node 30 is shared by the five elements 63, 64, 65, 66, and 67 (see

DISCRETIZATION WITH THE FINITE ELEMENT METHOD

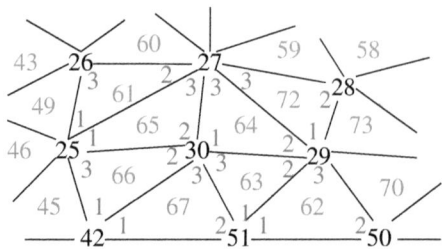

Figure 5.3: Part of a mesh with finite elements and grid nodes.

Fig. 5.3), which all make a contribution to the node 30. So the global matrix components $A^g_{30,j}$ are found with the help of index transformations by the way

$$A^g_{30,25} = A^{65}_{2,1} + A^{66}_{2,3}, \qquad A^g_{30,29} = A^{63}_{3,2} + A^{64}_{1,2}, \qquad \ldots \tag{5.94}$$

The assembling of the global matrix from all N elements with the index transformation $T(\mathbf{A}^e)$ can be described in the form

$$\mathbf{A}^g = \sum_{e=1}^{N} T(\mathbf{A}^e). \tag{5.95}$$

Sometimes there are more than one variable on the grid nodes. For example, with two variables the size of the local matrix $\mathbf{A}^e$ (two dimensions) is 6×6 and $2N \times 2N$ for the global matrix $\mathbf{A}^g$. If the variables are independent, the offset of the entries for the second variable is 3 in $\mathbf{A}^e$ and N in $\mathbf{A}^g$. For assembling the second variable from the element 66 the index transformation (5.93) must be modified by adding the respective offset

$$A^{66}_{3+1,3+1} \to A^g_{N+42,N+42}, \quad A^{66}_{3+1,3+2} \to A^g_{N+42,N+30}, \quad A^{66}_{3+1,3+3} \to A^g_{N+42,N+25}, \ldots \tag{5.96}$$

5.3.2 Dirichlet Boundary Conditions

Through the Dirichlet boundary conditions the values on the surface grid nodes are already fixed with the so-called Dirichlet value. Therefore, it is not necessary and even not allowed to recalculate the values on these grid nodes from the global equation system $\mathbf{A}^g \, \varphi^g = b^g$, because it is impossible to obtain the same Dirichlet values by solving the equation system. These surface grid nodes must be treated differently with the Dirichlet value. If on the global node i there is a Dirichlet boundary value $\varphi_i = C_i$, the global equation system

5.3 Assembling and Solving

must be changed to

$$\begin{bmatrix} a_{1,1} & a_{1,2} & \cdots & a_{1,i-1} & 0 & a_{1,i+1} & \cdots & a_{1,N} \\ a_{2,1} & a_{2,2} & \cdots & a_{2,i-1} & 0 & a_{2,i+1} & \cdots & a_{2,N} \\ \vdots & \vdots & \vdots & \vdots & \vdots & \vdots & \vdots & \vdots \\ a_{i-1,1} & a_{i-1,2} & \cdots & a_{i-1,i-1} & 0 & a_{i-1,i+1} & \cdots & a_{i-1,N} \\ 0 & 0 & \cdots & 0 & 1 & 0 & \cdots & 0 \\ a_{i+1,1} & a_{i+,2} & \cdots & a_{i+1,i-1} & 0 & a_{i+1,i+1} & \cdots & a_{i+1,N} \\ \vdots & \vdots & \vdots & \vdots & \vdots & \vdots & \vdots & \vdots \\ a_{N,1} & a_{N,2} & \cdots & a_{N,i-1} & 0 & a_{N,i+1} & \cdots & a_{N,N} \end{bmatrix} \begin{bmatrix} \varphi_1 \\ \varphi_2 \\ \vdots \\ \varphi_{i-1} \\ \varphi_i \\ \varphi_{i+1} \\ \vdots \\ \varphi_N \end{bmatrix} = \begin{bmatrix} b_1 \\ b_2 \\ \vdots \\ b_{i-1} \\ C_i \\ b_{i+1} \\ \vdots \\ b_N \end{bmatrix}. \quad (5.97)$$

From the mathematical point of view the global equation system $\mathbf{A^g}\,\varphi^g = b^g$ has m pseudo-equations if there are m grid nodes with Dirichlet conditions, after setting all m rows and columns from Dirichlet grid nodes φ_i in $\mathbf{A^g}$ to 0.

In practice it is more comfortable to multiplicate $\mathbf{A^g}$ with a transformation matrix $\mathbf{T_b}$

$$\mathbf{T_b}\,\mathbf{A^g}\,\varphi^g = b^g, \quad (5.98)$$

which sets all rows and columns for the m Dirichlet grid nodes $\varphi_i = C_i$ in $\mathbf{A^g}$ to 0, instead of doing it componentwise by $A^g_{ik} = 0$ and $A^g_{ki} = 0$ for $k = 1\ldots N$. In the beginning $\mathbf{T_b}$ is a unit matrix ($T_{ii} = 1$ and $T_{ij} = 0$), but for every Dirichlet grid node i the components T_{ii} are reset to $T_{ii} = 0$. Therefore, all m rows and columns in $\mathbf{A^g}$ can easily be set to 0 at once with $\mathbf{T_b}$.

5.3.3 Mechanical Interfaces

The simulated structures generally consist of several segments. Normally a segment is a continuous region of one material and so there are everywhere the same material characteristics. Therefore, the electrical and mechanical behavior can be described with the same parameters within a segment.

The used meshing module makes a separated grid for every segment. This means that for a global grid point at the interface between two different segments there exist two different indices, because each segment has its own global index, as shown in Fig. 5.4. The two global indices for the same grid point lead to an important aspect for the mechanics with regard to finite elements, because in the global stiffness matrix $\mathbf{K^g}$ there exist entries for two node indices for the same grid point. After solving the linear equation system $\mathbf{K^g}\,\vec{d}^g = \vec{f}^g$ for the mechanical problem there would be two different, but wrong displacement vectors for the same grid point.

The calculated displacements would be wrong, because the entries from segment A for this interface points in the global stiffness matrix do not take into account that there is

DISCRETIZATION WITH THE FINITE ELEMENT METHOD

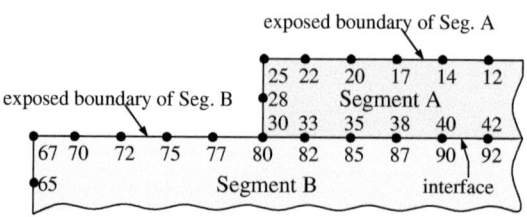

Figure 5.4: Two segments with interface and its grid points.

also a stiffness from the other segment B due to the separate assembling of the segments. The same but reverse explanation is valid for the entries from segment B. Therefore, the stiffness in matrix $\mathbf{K^g}$ and also the force in $\vec{f}^g$ must be corrected in the way that the entries with second index are added to the first index of the grid point and then all entries with second index are set to 0.

In the following the correcting procedure in $\mathbf{K^g}$ is demonstrated for the representative interface point with Index 35 in Segment A and Index 85 in Segment B (see Fig. 5.4).

$$\mathbf{K^g} = \begin{bmatrix} k_{1,1} & \cdots & k_{1,34} & k_{1,35}+k_{1,85} & k_{1,36} & \cdots & k_{1,N} \\ k_{2,1} & \cdots & k_{2,34} & k_{2,35}+k_{2,85} & k_{2,36} & \cdots & k_{2,N} \\ \vdots & \vdots & \vdots & \vdots & \vdots & \vdots & \vdots \\ k_{34,1} & \cdots & k_{34,34} & k_{34,35}+k_{34,85} & k_{34,36} & \cdots & k_{34,N} \\ k_{35,1}+k_{85,1} & \cdots & k_{35,34}+k_{85,34} & k_{SUM} & k_{35,36}+k_{85,36} & \cdots & k_{35,N}+k_{85,N} \\ k_{36,1} & \cdots & k_{36,34} & k_{36,35}+k_{36,85} & k_{36,36} & \cdots & k_{36,N} \\ \vdots & \vdots & \vdots & \vdots & \vdots & \vdots & \vdots \\ k_{N,1} & \cdots & k_{N,34} & k_{N,35}+k_{N,85} & k_{N,36} & \cdots & k_{N,N} \end{bmatrix}$$

with $\quad k_{SUM} = k_{35,35} + k_{35,85} + k_{85,35} + k_{85,85}$. (5.99)

After the addition of all entries with Index 85 these components are set to zero: $k_{i,85} = 0$ and $k_{85,i} = 0$ for $i = 1\ldots N$. The other interface points are handled in the same way. It is also necessary to rearrange the force vector $\vec{f}^g$ with the same concept so that

$$\vec{f}^g = \begin{bmatrix} f_1 & f_2 & \cdots & f_{34} & f_{35}+f_{85} & f_{36} & \cdots & f_{84} & 0 & f_{86} & \cdots & f_N \end{bmatrix}^T. \quad (5.100)$$

Like in (5.98) $\mathbf{K^g}$ and $\vec{f}^g$ can be manipulated for all interface points in a faster and simpler way with the same transformation matrix $\mathbf{T_b}$, because the Dirichlet boundary conditions have higher priority and must always be fulfilled. With $\mathbf{T_b}$ the mechanical system becomes

$$\mathbf{T_b} \mathbf{K^g} \vec{d}^g = \mathbf{T_b} \vec{f}^g. \quad (5.101)$$

For the representative interface point with Index 35/85 the matrix elements from $\mathbf{T_b}$ must be set to $T_{35,85} = 1$ and $T_{85,85} = 0$ in order to get the required effect.

After solving the linear system for displacements (5.101), d_{35} has the correct value and $d_{85} = 0$. In order to also have the correct displacement for the second Index 85 (segment B), d_{85} must be set to d_{35}.

5.3.4 Complete Equation System for Oxidation

The oxidation problem induces 5 unknown variables: the oxidation concentration c, the normalized silicon concentration η, and the three components u, v, and w (in x-, y-, and z-direction) of the displacement vector $\vec{d}$. If the dopant redistribution is taken into account, than the unknown variables for the concentrations of the different species have also to be considered. In case of the five-stream diffusion model these are the 5 variables for C_{A^+}, C_I, C_V, C_{AI}, and C_{AV} (see Section 4.2).

The following explanation is focused on the pure oxidation problem. Therefore, the discretization of the simulation domain with N grid nodes leads to 5 variables on each grid node. After assembling of the whole equation system there are in total $5N$ unknown variables and, as already described in Section 5.3.1, the dimension of the global matrix $\mathbf{A^g}$ is $5N \times 5N$.

As shown in Fig. 5.5 the global matrix $\mathbf{A^g}$ consists of the two sub-matrices $\mathbf{J^g}$ and $\mathbf{K^g}$. $\mathbf{J^g}$ contains the coupled entries from the oxidant diffusion (5.64) and the η-dynamics (5.73). Because of the unknown variables $c_1 \ldots c_N$ and $\eta_1 \ldots \eta_N$ the size of $\mathbf{J^g}$ is $2N \times 2N$. $\mathbf{K^g}$ is the global stiffness matrix for the mechanics with the three displacement components $u_1 \ldots u_N$, $v_1 \ldots v_N$, and $w_1 \ldots w_N$, and so its dimension is $3N \times 3N$.

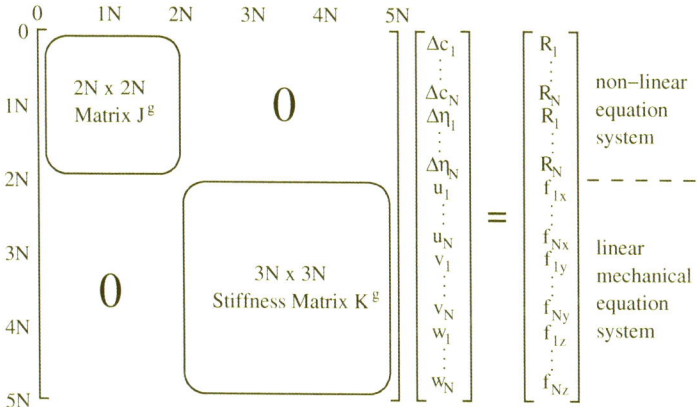

Figure 5.5: Structure of the global equation system for oxidation.

DISCRETIZATION WITH THE FINITE ELEMENT METHOD

The structure of the local system matrix for one finite element is the same as for the global one (see Fig. 5.5). The only difference is that $N = 4$ because of the four nodes of a tetrahedron. So the dimension of the local matrix $\mathbf{A^e}$ is 20×20 with the 8×8 $\mathbf{J^e}$ and 12×12 $\mathbf{K^e}$ sub-matrices.

The first part of the equation system which describes the oxidation diffusion and the η-dynamics, is non-linear because of the coupling between c_i and η_i in the form $c_i \eta_i$ (see (5.64) and (5.73)). The variables in the non-linear system can not be calculated directly, only their increments Δc_i and $\Delta \eta_i$. The second part of the system which is responsible for the mechanical problem is linear, as marked in Fig. 5.5.

For a quasi-stationary time step there is no coupling between the c-η-system and the mechanical system, because the equations for c and η are not functions of displacements. Although the normalized additional volume after a time step is calculated with c and η, the mechanical equations for the displacements also are not functions of c and η. Because of the missing coupling the off-diagonal sub-matrices in the global equation system are zero (see Fig. 5.5).

On the right-hand side of the non-linear subsystem are the residuals calculated with the results from the last Newton iteration as explained in the next section. The right-hand side of the linear subsystem contains the (internal) forces on the grid nodes.

5.3.5 Solving with the Newton Method

In contrast to the linear mechanical subsystem, where the displacements can be calculated directly with one of the various standard methods like Gaussian elimination, the solving of the non-linear part demands other routines like the used Newton Method [96].

The global non-linear subsystem can be written in the form

$$\begin{aligned} f_1(x_1, x_2, \ldots, x_{2N}) &= 0 \\ f_2(x_1, x_2, \ldots, x_{2N}) &= 0 \\ &\vdots \\ f_{2N}(x_1, x_2, \ldots, x_{2N}) &= 0 \end{aligned} \tag{5.102}$$

where $x_1 \ldots x_N$ stand for the variables $c_1 \ldots c_N$ and $x_{N+1} \ldots x_{2N}$ stands for the variables $\eta_1 \ldots \eta_N$.

With the Newton formula the solution vector for the actual time step n becomes

$$\vec{x}^n = \vec{x}^{n-1} - \mathbf{J^{g}}^{-1}(x^{n-1}) \vec{R}(x^{n-1}), \tag{5.103}$$

where $\vec{x}^{n-1}$ is the solution from the previous time step $n-1$, $\mathbf{J^{g}}^{-1}$ is the inverse Jacobian matrix, and $\vec{R}$ are the residuals, both determined by x_i^{n-1}.

5.3 Assembling and Solving

Transforming the Newton formula to the form

$$\mathbf{J^g}\vec{\Delta x} = \vec{R} \qquad \text{with} \qquad \Delta x_i = x_i^n - x_i^{n-1}, \tag{5.104}$$

leads to a linear equation system for the increments Δx_i (see Fig. 5.5).

After solving this linear system the values for the actual timestep n can be determined in the way

$$x_i^n = x_i^{n-1} + \Delta x_i. \tag{5.105}$$

This equation shows that the Newton method demands start values x_i^0 on all N grid nodes for the first iteration $n = 1$.

Because the Newton method is only a first order approximation of the solution, an iteration never provides the exact results. Therefore the right-hand side of the non-linear system (5.102) can not be 0, instead there always exist residuals $R_i(x_i^{n-1})$.

The quality of the approximation is increased with each iteration, but the number of iterations must be limited with termination conditions. With these conditions also the accuracy of the approximation can be controlled. It makes sense to use the following termination conditions [97]

$$\|\vec{x}^n - \vec{x}^{n-1}\| \leq \tau_{abs}, \tag{5.106}$$

$$\|\vec{x}^n - \vec{x}^{n-1}\| \leq \tau_{rel}\|\vec{x}^n\|, \tag{5.107}$$

$$\|\vec{R}(x^n)\| \leq \tau_f, \tag{5.108}$$

where τ_{abs}, τ_{rel}, and τ_f are given tolerances. $\|\ \|$ is the Euclidean norm

$$\|\vec{x}^n - \vec{x}^{n-1}\| = \sqrt{\sum \left(x_i^n - x_i^{n-1}\right)^2} \qquad \text{and} \qquad \|\vec{x}^n\| = \sqrt{\sum \left(x_i^n\right)^2}. \tag{5.109}$$

(5.106) and (5.107) are failure criterions, which only work, if the sequence $\vec{x}^n$ converges quickly to the exact solution. If this converging sequence is slow, $\|\vec{x}^n - \vec{x}^{n-1}\|$ can be small but the demanded accuracy is by far not reached and the Newton loop is terminated too early. Due to this fact it is recommended to use the additional residual criterion (5.108). The combination of both criterions ensures good terminating conditions.

For one finite tetrahedral element with its four nodes ($N = 4$) the non-linear system (5.102) is built up as

$$\begin{aligned} f_1(c_1^n, c_2^n, c_3^n, c_4^n, \eta_1^n, \eta_2^n, \eta_3^n, \eta_4^n) &= 0, \\ f_2(c_1^n, c_2^n, c_3^n, c_4^n, \eta_1^n, \eta_2^n, \eta_3^n, \eta_4^n) &= 0, \\ &\vdots \\ f_8(c_1^n, c_2^n, c_3^n, c_4^n, \eta_1^n, \eta_2^n, \eta_3^n, \eta_4^n) &= 0. \end{aligned} \tag{5.110}$$

DISCRETIZATION WITH THE FINITE ELEMENT METHOD

Here $f_1 \ldots f_4$ come from the oxidant diffusion (5.64)

$$f_j = \sum_{i=1}^{4} \left(D K_{ij} c_i^n + k_{max} M_{ij} c_i^n \eta_i^n \right) = 0 \quad \text{for} \quad j = 1, 2, 3, 4, \tag{5.111}$$

and the other equations $f_5 \ldots f_8$ describe the η-dynamics (5.73)

$$f_{j+4} = \sum_{i=1}^{4} \left(M_{ij}(\eta_i^n - \eta_i^{n-1}) + K_A M_{ij} \eta_i^n c_i^n \Delta t \right) = 0, \quad \text{for} \quad j = 1, 2, 3, 4. \tag{5.112}$$

The local Jacobian matrix is

$$\mathbf{J^e} = \begin{bmatrix} \frac{\partial f_1}{\partial c_1^n} & \frac{\partial f_1}{\partial c_2^n} & \frac{\partial f_1}{\partial c_3^n} & \frac{\partial f_1}{\partial c_4^n} & \frac{\partial f_1}{\partial \eta_1^n} & \frac{\partial f_1}{\partial \eta_2^n} & \frac{\partial f_1}{\partial \eta_3^n} & \frac{\partial f_1}{\partial \eta_4^n} \\ \frac{\partial f_2}{\partial c_1^n} & \frac{\partial f_2}{\partial c_2^n} & \frac{\partial f_2}{\partial c_3^n} & \frac{\partial f_2}{\partial c_4^n} & \frac{\partial f_2}{\partial \eta_1^n} & \frac{\partial f_2}{\partial \eta_2^n} & \frac{\partial f_2}{\partial \eta_3^n} & \frac{\partial f_2}{\partial \eta_4^n} \\ \vdots & \vdots & \vdots & \vdots & \vdots & \vdots & \vdots & \vdots \\ \frac{\partial f_8}{\partial c_1^n} & \frac{\partial f_8}{\partial c_2^n} & \frac{\partial f_8}{\partial c_3^n} & \frac{\partial f_8}{\partial c_4^n} & \frac{\partial f_8}{\partial \eta_1^n} & \frac{\partial f_8}{\partial \eta_2^n} & \frac{\partial f_8}{\partial \eta_3^n} & \frac{\partial f_8}{\partial \eta_4^n} \end{bmatrix}. \tag{5.113}$$

For calculating the Jacobian matrix and the residuals in the Newton iteration n the results for c_i and η_i from the previous iteration $n-1$ are used.
For the equations $f_1 \ldots f_4$ (5.111) of the Jacobian matrix components are

$$J_{i,j} = \frac{\partial f_i}{\partial c_j^n} = D K_{ij} + k_{max} M_{ij} \eta_i^{n-1},$$

$$J_{i,j+4} = \frac{\partial f_i}{\partial \eta_j^n} = k_{max} M_{ij} c_i^{n-1} \quad \text{for} \quad i, j = 1, 2, 3, 4. \tag{5.114}$$

The partial derivatives of the equations $f_5 \ldots f_8$ (5.112) result in

$$J_{i+4,j} = \frac{\partial f_{i+4}}{\partial c_j^n} = K_A M_{ij} \eta_i^{n-1} \Delta t,$$

$$J_{i+4,j+4} = \frac{\partial f_{i+4}}{\partial \eta_j^n} = M_{ij} + K_A M_{ij} c_i^{n-1} \quad \text{for} \quad i, j = 1, 2, 3, 4. \tag{5.115}$$

The local residuals used for the actual Newton iteration n are

$$R_i = \sum_{j=1}^{4} \left(D K_{ij} c_i^{n-1} + k_{max} M_{ij} c_i^{n-1} \eta_i^{n-1} \right), \tag{5.116}$$

$$R_{i+4} = \sum_{j=1}^{4} \left(M_{ij}(\eta_i^{n-1} - \eta_i^{n-2}) + K_A M_{ij} \eta_i^{n-1} c_i^{n-1} \Delta t \right) \quad \text{for} \quad i = 1, 2, 3, 4. \tag{5.117}$$

The entries of the global Jacobian matrix $\mathbf{J^g}$ and global residual vector $\vec{R}$ as needed for (5.104) are assembled from the local ones as explained in Section 5.3.1.

Chapter 6

Simulation of Thermal Oxidation with FEDOS

FEDOS stands for **F**inite **E**lement **D**iffusion and **O**xidation **S**imulator and is in principle a framework for three-dimensional process simulation, which is based on the finite element method. The name has more traditional character and does not enumerate all its abilities, because, when FEDOS was launched, it was only planned to simulate different forms of diffusion and thermal oxidation processes. Since the concept of FEDOS allows to simulate all process phenomena, if the problem can be formulated with the finite element method, it is also used for the investigation of other process topics like electromigration or stress analysis. In the course of this doctoral work FEDOS was extended and modified for the simulation of oxidation and various kinds of stress analysis.

The finite element method offers some benefits in process simulation compared with other numerical techniques. At first it enables to discretize all kinds of (partial differential) equations in a similar way and with good mathematical stability. Because FEM was developed for mechanical simulation, it is also most suitable for displacement problems as occur during thermal oxidation. Another advantage is that after the discretization of the equations which describe the respective physical phenomenon analytically, FEM only needs standardized routines to built up the global equation system. This means that in FEDOS the same assembling procedure can be used for all different process models.

The FEM formulation goes hand in hand with the used elements. In the current version FEDOS is designed for simulation regions which are exclusively discretized with tetrahedrons and linear shape functions. Since the accuracy can be increased with a finer mesh, which means more elements, the linear FEM approach meets all requirements and has the advantage that it is the most simple FEM formulation (see Section 5.2). Furthermore, tetrahedrons are qualified for fitting non-planar surfaces with coarse elements in acceptable quality.

SIMULATION OF THERMAL OXIDATION WITH FEDOS

Regarding the implementation aspect an advantage of FEDOS is that a new model can be included in a straightforward procedure. It is only necessary that the new model is implemented in C++ with a defined interface in a separate file. For including a model in FEDOS only a knowledge about the program interface is demanded, but not about the complex internal FEDOS routines or even other models. The new model must only supply the finite element formulation of the discretized equations which describe the phenomena on a single element.

6.1 Architecture of FEDOS

The core functions of FEDOS are the management of the simulation procedure and data flow, the model execution, and the finite element assembling. Some functions like data- and Inputdeck-file operations or the solving of the equation system are provided by libraries, but these function calls are incumbent on FEDOS. Furthermore, FEDOS offers a number of operations for mesh manipulation, especially for (dynamic) mesh refinement and coarsement.

6.1.1 Inputdeck

FEDOS always asks for a so-called Inputdeck-file (ipd-file) which includes all necessary information for a simulation run. In principle the ipd-file contains all changeable process information. The Inputdeck-file can be read with the Inputdeck-Reader which is a library linked into FEDOS. The Inputdeck concept was also developed at the Institute for Microelectronics (see Chapter 3 in [98]) and is also used for other simulators. The ipd-file itself is an Ascii-file which can be generated with a normal text editor in an evident syntax.

The ipd-file includes the names of the input and the output file. Alternatively sometimes it is desired to set an attribute to a constant initial value on the whole segment. In the case of oxidation simulation the normalized silicon concentration η must be set to the initial value 1 in the silicon segment (see Section 3.1). The next important task is to control the simulation procedure which involves amongst others the

- Maximal time of the simulated process
- Duration of one time step
- Number of time steps
- If the duration of the time steps is constant or increased recursively

In the ipd-file also the accuracy for the Newton solver is determined (see Section 5.3.5) which is related with the controlling part. For a desired higher accuracy of the results more Newton loops and so more simulation time is needed for solving a non-linear equation system.

6.1 Architecture of FEDOS

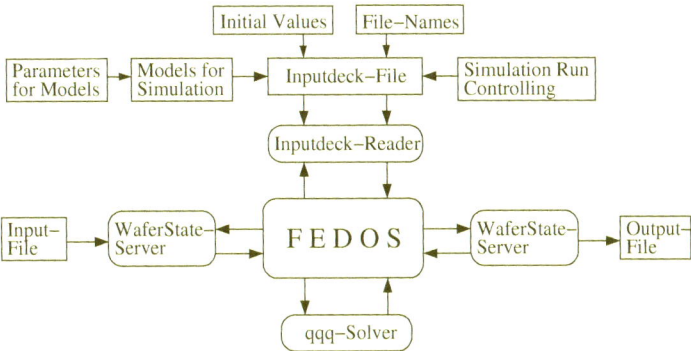

Figure 6.1: Architecture of FEDOS and its information flow.

Since FEDOS contains a number of different models, another necessary part in the ipd-file is to declare which model is applied on the respective segment by its name. The models can be divided into the three categories:

- Volume models: describe the physical behavior within a segment. Examples are the models for oxidation, diffusion, or mechanics. Furthermore, it is obligatory to assign a volume model to each segment. This means that also on not relevant segments of a more complex simulation setup a dummy model must be applied.

- Surface models: can be applied on segment surfaces with boundary conditions. The surface models always contain a Dirichlet or Neumann boundary condition. Boundary conditions for the mechanics (Dirichlet) or the species flow for diffusion (Neumann) can be listed. On surfaces without explicit models FEM assumes implicitly a Neumann boundary condition which means that there exists no flow of particles through the surface. In contrast to the volume models the mathematical formulation in the surface model has to be performed for triangles, because a surface only has two dimensions.

- Interface models: describe the physical behavior on the interface between two adjacent segments. Like the surface also an interface only has two dimensions. As depicted in Section 5.3.3 an interface model is essential for mechanical problems. Another example is the segregation of species at an interface.

Simulation only makes sense, when the process parameters are changeable and so at least all volume models have its own parameters. For the oxidation model such parameters are the low stress diffusion coefficient D_0 (see (3.2)) or the maximal strength of the spatial sink k_{max} (see (3.4)). In the mechanical models the Young modulus E and the Poisson ratio ν are modifiable.

SIMULATION OF THERMAL OXIDATION WITH FEDOS

6.1.2 Wafer-State-Server

For the data management FEDOS uses the WAFER-STATE-SERVER [99], a program package developed at the Institute for Microelectronics. All data are saved in the so-called WAFER-STATE-SERVER-file (WSS-file) in an Ascii-format. The WSS-format enables straightforward communication of FEDOS with other in-house tools as for meshing of the structure or visualization of the simulation results.

In the WSS-file are one or more segments where each segment holds a (tetrahedral) grid. On the segment grid a unlimited number of constant or distributed attributes can be located. The WSS-file concept has the benefit regarding the file size that the coordinates (x-, y- and z-value) of each grid point are only saved once although, a grid point is shared by a number of tetrahedrons. Therefore, the nodes of the tetrahedrons in the segment grid are only references to a point list. Another memory saving effect is that the distributed attribute values are also saved only once on the grid points in the respective segment and not on each tetrahedron node.

The WAFER-STATE-SERVER is not merely a file reading and writing tool, it is in principle a data management tool. In the beginning all grid and attribute information from the input file are read and then held in the WAFER-STATE-SERVER during the simulation. It achieves an abstraction of the physical stored data in the file to logical dats in the program. For FEDOS the WAFER-STATE-SERVER supplies a lot of useful grid operations like surface and interface extraction, point and element location, or attribute updates during the simulation. For simulation with FEDOS the WSS input file must at least contain the grid information of the discretized structure. The simulation results are written to the output file in form of distributed attributes. For the oxidation simulation the results are the distribution of the oxidant concentration C and the normalized silicon concentration η. For the mechanical problem with its displacements also the point coordinates are modified in the output file.

6.1.3 QQQ-solver

The solving of the (linear) global equation system is performed with the QQQ-solver [100], also developed at the Institute for Microelectronics. The QQQ-solver is based on the Gaussian method [101], which uses a factorization of the matrix $\mathbf{A^g}$ in a lower and upper triangular matrix $(\mathbf{A} = \mathbf{L} \cdot \mathbf{U})$, so that the equation system $\mathbf{A} \cdot \vec{x} = \vec{b}$ can be written as

$$\mathbf{L} \cdot \mathbf{U} \cdot \vec{x} = \vec{b}, \qquad \mathbf{L} \cdot \vec{y} = \vec{b} \quad \text{with} \quad \vec{y} = \mathbf{U} \cdot \vec{x}. \tag{6.1}$$

Therefore, the Gaussian algorithm is specified by the following three steps:

1. $\mathbf{A} = \mathbf{L} \cdot \mathbf{U}$: Gaussian elimination by factorization ($\mathbf{L}$ and $\mathbf{U}$ is computed)
2. $\mathbf{L} \cdot \vec{y} = \vec{b}$: forward-substitution ($\vec{y}$ is computed)
3. $\vec{y} = \mathbf{U} \cdot \vec{x}$: backward-substitution ($\vec{x}$ is computed)

The QQQ-solver also supplies a transformation matrix $\mathbf{T_b}$ which allows to transform the equation system $\mathbf{A} \cdot \vec{x} = \vec{b}$ to [102]

$$\mathbf{T_b} \cdot \mathbf{A} \cdot \vec{x} = \mathbf{T_b} \cdot \vec{b}. \tag{6.2}$$

As depicted in Section 5.3.2 and 5.3.3 the matrix $\mathbf{T_b}$ can be used for the elimination of equations not needed because of Dirichlet boundary conditions or for correcting the equation system in case of mechanical interfaces.

The assembling of the equation system is performed by FEDOS by generating the matrices $\mathbf{A}$, $\mathbf{T_b}$ and $\vec{b}$ for the QQQ-module. After solving the QQQ-module returns the results to FEDOS. The complete equation system for the oxidation problem (see Section 5.3.4) consists of the non-linear (diffusion-reaction) part and the linear (mechanical) part.

The non-linear sub-system requires some Newton iterations, until it fulfills the termination conditions. It should be mentioned that the QQQ-module is not a non-linear solver, it can only can handle linear systems. As described in Section 5.3.5 FEDOS assembles the non-linear sub-system in such a kind that it becomes a linear system for increments Δx_i, which can be solved by the QQQ-module. These increments are computed in a way that FEDOS can build a solution. This procedure is repeated until the approximation fulfills the desired terminating conditions.

6.2 Simulation Procedure

The first step of the simulation procedure is to perform a finite element discretization by splitting up the three-dimensional structure into tetrahedral elements. A key aspect for the simulation is the number of elements, because it determines the accuracy of the simulation results and the demanded computer resources. A finer mesh with more elements means that the larger number of nodes leads to a larger equation system which needs more time for its assembling and solving procedure. The mesh generation is performed with the meshing tool which results in the input data file for FEDOS. Then the ipd-file with all the simulation parameters is imported with the Inputdeck-Reader and the file which contains the mesh, geometry, and material information is read into the WAFER-STATE-SERVER.

In the next step the initial values for the oxidant concentration C and the normalized silicon concentration η are set on the grid nodes. For example η must be 1 in pure silicon. Because the oxidation process is time dependent, the actual oxidation time must be reset at the beginning of the simulation. As shown in Fig. 6.2, FEDOS iterates over all finite elements and builds the local equation system for every element at each actual discrete time. The local equation system describes the oxidation process numerically only for one element. In order to describe the whole oxidation process on the complete simulation domain the finite element method demands a global (coupled) equation system. The

SIMULATION OF THERMAL OXIDATION WITH FEDOS

components of the global equation system are assembled from the local system by using the superposition principle as depicted in Section 5.3.1.

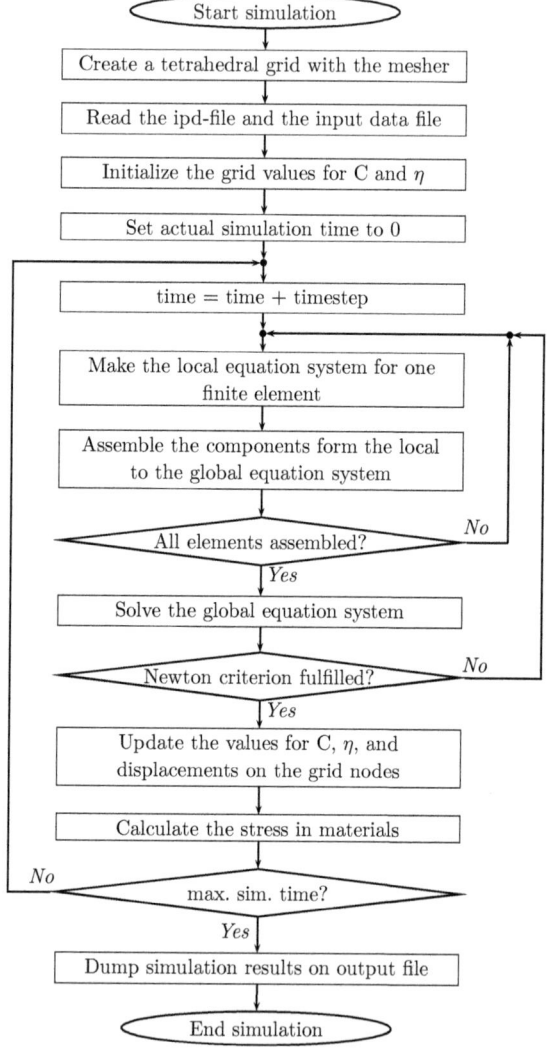

Figure 6.2: Simulation procedure for oxidation.

After the iteration over all elements is finished, the global assembled equation system, with its non-linear and linear part, is also completed. Now the global equation system can be solved with the QQQ-solver. The assembling and solving procedure is repeated, until the results from the non-linear sub-system fulfill the termination conditions of the Newton method. After the Newton system has converged the results for C, η, and displacements for the whole discretized oxidation process are obtained for the actual time step.

With these results the values for C, η, and the displacement are updated on the grid nodes such that these values are always keeping pace with the actual simulation time. The actual displacement vector enables the calculation of the strain tensor as well as the stress tensor for each element. When the above described procedure is finished, the actual simulation time is increased and the assembling for the first Newton loop is started again. The same assembling and solving procedure is repeated for each time step, until the desired end of the simulation. At the end of the simulation procedure the WAFER-STATE-SERVER writes the final simulation results to the output file.

6.3 Meshing Aspects

The mesh generation for the (oxidation) simulation with FEDOS is an important topic, because the finite element formulation depends on the used elements. With the actual FEDOS version the discretization can be performed only with linear shape functions on tetrahedrons. This tetrahedral grid generation is performed with the in-house meshing tool LAYGRID from the *Smart Analysis Programs* package [103]. The quality of the numerical solution of the PDEs by the finite element method increases with the number of nodes.

For a desired high accuracy of the simulation results a fine mesh with a high number of elements and nodes is requested [104]. If the mesh is not fine enough, there is a risk that the Newton method does not converge for the discretized non-linear equation system because of too large approximation failure. On the other side a large number of elements and nodes has an unwanted effect: more computer resources are required, because it must be iterated over more elements which also must be assembled to the global equation system. In case of the oxidation model there are five variables on each node, so that an additional node results in five additional equations. A larger equation system needs more time and memory for its solving. Therefore, the goal for finite elements is always to obtain a high accuracy with the smallest possible number of elements.

For the oxidation simulation a static grid is used which has the advantage that grid manipulation procedures are not needed. A grid modification like refining and coarsening in each time step normally needs complex algorithms with a long computation time and has the risk of element degeneration [105]. So the best way to reach the above goal with a static grid is to make an initial mesh with appropriate local resolution. In critical or intersting regions of the investigated structure, or where the oxidation process really occurs, a finer mesh should be applied than in the rest of the structure.

The previously discussed meshing strategy is applied to discretize an initial structure as shown in Fig. 6.3. This demonstrative example is a silicon block with (1.2×0.3) μm floor space and a height of 0.4 μm. Two thirds of the length are covered with a 0.15 μm thick silicon nitride mask which prevents the oxidant diffusion on the subjacent silicon block. In principle this chosen structure is two-dimensional, but it is very suitable for the plausible illustration how the sharp interface interpretation (see Section 6.4) and the stress calculation strategy (see Section 7.2) works. The first interesting information regarding meshing is that the oxidation process only starts at the upper uncovered silicon surface. The next important aspect is that the most critical region on this structure is along the edge of the Si_3N_4-mask. This area is of interest, because the stiffness of the Si_3N_4-mask prevents the desired volume expansion of the newly formed oxide, which leads to the well known bird's beak effect.

Therefore, the finest mesh in the structure was constructed around the mask edge. The distance of the nodes in the x-direction is 10 nm. For nodes which are located away located from the edge their distance is successively increased until 50 nm at the end of the active silicon region (x = 0 μm). At the end of the Si_3N_4-mask (x = 1.2 μm) the node distance in x-direction is even 100 nm. Furthermore, on the upper half of the silicon block where the oxidation process is expected the layer thickness is 10 nm, in the lower half it is doubled (20 nm). Unfortunately LAYGRID is limited to produce only layers with constant thickness over the whole x-y-plane. Therefore, the layer thickness in the less interesting regions somewhere under the mask must be the same as in the active area. All in all, the mesh shown in Fig. 6.3 has 12 218 nodes and 56 670 (tetrahedral) elements.

The results of the simulated oxidation process, which are the η-distribution and the displacements in the materials, are displayed in Fig. 6.4. Here, blue is pure SiO_2 ($\eta = 0$), red is the pure silicon substrate ($\eta = 1$), and at the Si/SiO_2-interface one can see the reaction layer with a finite width ($0 < \eta < 1$) as explained in Section 3.1. Furthermore, this figure depicts the node displacements and the grid deformation caused by the considerable volume increase of the newly formed SiO_2.

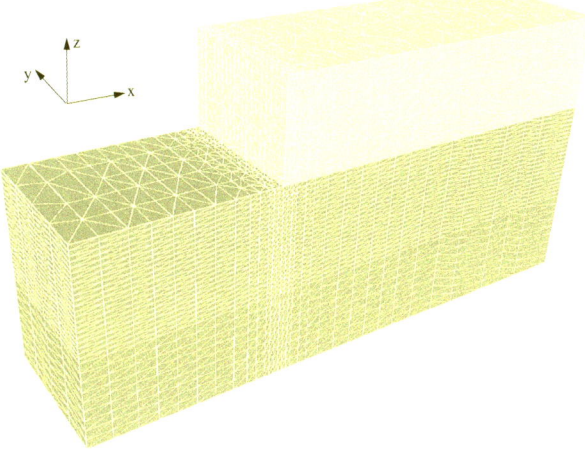

Figure 6.3: Tetrahedral mesh with different fineness on the initial structure.

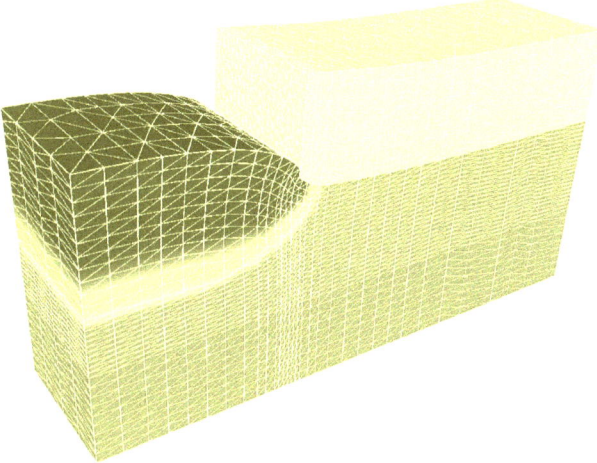

Figure 6.4: Simulation result of the oxidation process with grid deformation.

6.4 Sharp Interface and Smoothing

Since the η-distribution is only a virtual model parameter, the width of the reaction layer does not agree with the thickness of the real physical interface between silicon and SiO_2. In the calculated η-distribution the reaction layer normally ranges over some finite elements (see Fig. 6.4), but in reality the Si/SiO_2-interface is only a few atom layers thick. So for a more physical presentation of the (final) simulation result a sharp interface between silicon and SiO_2 must be constructed.

6.4.1 Segment Splitting

The two regions can be extracted from the η-distribution by determining that $\eta \leq 0.5$ is SiO_2 and $\eta > 0.5$ is silicon. From the meshing aspect this means that the original silicon segment must be splitted up into two new segments, one for pure silicon and another for pure SiO_2, which can be done by cutting the grid at a virtual surface with $\eta = 0.5$.

For the sake of simplicity the splitting procedure is demonstrated on a two-dimensional grid example. The simulated structures are three-dimensional with a tetrahedral mesh, but the principle is the same as with triangles. The left side of Fig. 6.5 shows a subarea of a mesh with the η-values on the nodes. There the η-values on the upper nodes are less than 0.5 and on the lower nodes are higher than 0.5. This means that the virtual surface with $\eta = 0.5$ must be located somewhere between the upper and lower nodes. The position of $\eta = 0.5$ on each element edge can be calculated with the known values η_1 and η_2 on the two corresponding nodes

$$\frac{|0.5 - \eta_1|}{|0.5 - \eta_2|} = \frac{l_1}{l_2} \qquad \text{where} \qquad l = l_1 + l_2. \tag{6.3}$$

l is the length of the element edge, and l_1 (l_2) is the distance between the location of $\eta = 0.5$ and the node with η_1 (η_2) along this edge. In Fig. 6.5 the location of the 0.5-line is presented with linear proportions, because its distances to the nodes were calculated with (6.3).

After the position of $\eta = 0.5$ is calculated on an edge, a new node is inserted and the edge is split into two parts. The new nodes are marked with red color in the right side of Fig. 6.5. With the help of additional nodes, which are not placed on the egdes, a local remeshing of the interface grid can be performed, which results in two separated segments for silicon and SiO_2 with a sharp interface. The mesh operations for this segment splitting were implemented in FEDOS.

Another problem associated with segment splitting is that the generated interface is not smooth, especially in critical regions or where it has a curvature. The reasons are numerical inaccuracies which come from the finite element discretization, but also from the

6.4 Sharp Interface and Smoothing

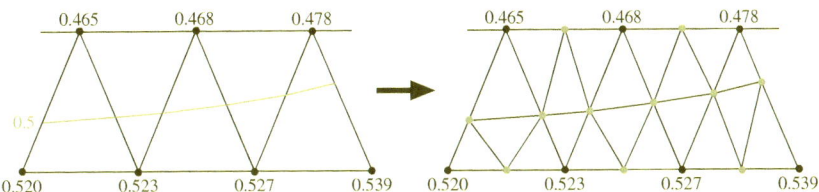

Figure 6.5: Principle of the grid operations for the splitting procedure.

Newton solving method because both are approximation methods. After a number of simulation loops (see Section 6.2) the inaccuracies sum up and lead to visible differences in the η-distribution.

The situation after the segment splitting of the oxidized structure from Fig. 6.4 is shown in Fig. 6.6. Only the silicon segment (with the same proportions) is presented. Although the mesh has good quality, the interface is craggy because of the previously described problems. For a more realistic Si/SiO_2-interface the quality of its curvature and mesh can be improved with an additional smoothing routine.

Figure 6.6: The Si/SiO_2-interface at the silicon segment after segment splitting.

SIMULATION OF THERMAL OXIDATION WITH FEDOS

6.4.2 Smoothing

The smoothing algorithm is implemented in the WAFER-STATE-SERVER in form of advanced GTS-functions [106]. The basic idea of the smoothing model is to move all points which are connected to artificial edges. An important part is to select which surface points belong to natural edges of the structure and which to artificial ones [107]. The principle of the point selection method can be explained with the help of Fig. 6.7. Points on planar surfaces like P_1 can be excluded from the smoothing process, because they are only surrounded by planar triangles. The same is valid for points like P_2, which are located on natural edges, because they are also connected with at least one planar surface.

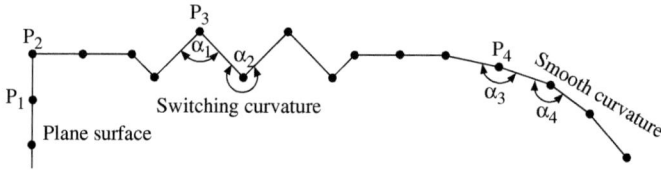

Figure 6.7: Principle of the point selection method for different kind of surfaces.

The best strategy for finding a point as P_3, which needs smoothing, is to check the surface curvature. A typical property of a point on an artificial edge is that the curvature of at least one connected other point is opposite. Such switching curvature can be located straightforwardly, with an angle criterion. As demonstrated in Fig. 6.7 the angle between the triangles at point P_3 is acute ($\alpha_1 < 180°$), but the angle at the connected point is obtuse ($\alpha_2 > 180°$). A plausible criterion for switching curvatures is to analyze, if the angles of connected points switch between less 180° and greater 180°. It can be found with this criterion that point P_4 belongs to an already smooth surface, because both angles α_3 and α_4 have similar values less than 180° ($\alpha_3, \alpha_4 < 180°$).

After selection of the points which have to move, their distances and directions of motion are another important aspects [107]. At first, the maximally allowed sphere of the motion of a point around its original position is given by the shortest distance to its connected points, as displayed in Fig. 6.8. Since the smoothing process is performed with a number of iterations, the distance of motion within each iteration loop is set to $\frac{1}{10}$ or less of the respective sphere radius. The direction of motion for a point for each iteration loop is calculated as the sum of normals of all triangles connected to this point (see right hand side in Fig. 6.8). The smoothing process for the selected points is stopped, if the difference of the angles between connected points is within a (small) tolerance

6.4 Sharp Interface and Smoothing

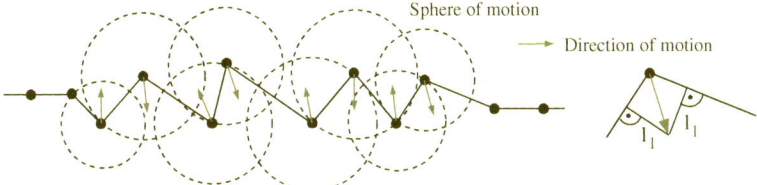

Figure 6.8: Illustration of the point motion concept in the smoothing process.

The above described method is applied to smoothen the Si/SiO$_2$-interface on the oxidized structure. The result of the smoothing processs for the silicon segment after approximately 20 iterations is shown in Fig. 6.9. It can be seen that compared with the interface after the segement splitting (see Fig. 6.6) the roughness of the smoothed interface is negligible because most artificial edges and unevennesses were removed.

Figure 6.9: The Si/SiO$_2$-interface at the silicon segment after the smoothing process.

The simulation results of the oxidation process after the previously described segment splitting and smoothing procedure (see Fig. 6.4), are presented with a more physical sharp interface between the SiO$_2$- and silicon segment in Fig. 6.10. It is worth mentioning that all pictures of this oxidation example have same proportions and perspectives for optimal comparison.

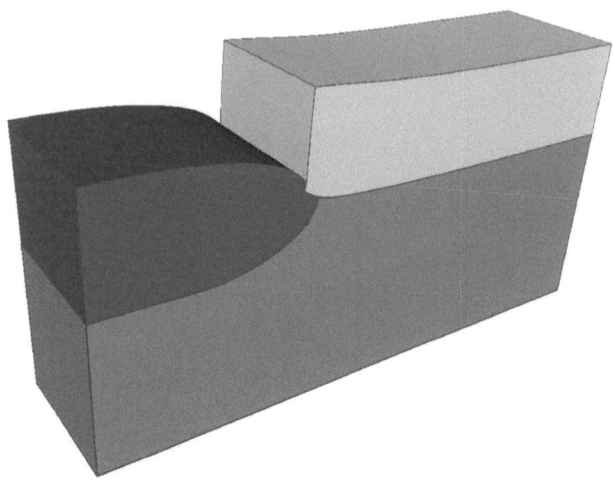

Figure 6.10: SiO$_2$-region after oxidation with a sharp and smoothed interface.

6.5 Model Calibration

The simulated oxide thickness after a certain oxidation time must agree with the real physical thickness under the same assumed process conditions. The goal was to find a universal, but not complicated calibration method which works for all possible oxidation conditions, as described in the following.

6.5.1 Calibration and Parameter Extraction

A look to the model (see Section 3.2.1) shows that there are three available parameters, namely the diffusion coefficient $D_0(T,p)$, the maximal possible strength of the spatial sink k_{max}, and the oxidant concentration C^{Sur} at surfaces which have contact to the oxidizing atmosphere. As displayed in (3.9) and (7.1) the diffusion coefficient has a physical background. It is temperature and stress dependent and its real physical value can be determined correctly. Therefore to use $D_0(T,p)$ for calibration is not appropriate.

The next parameter k_{max} has more mathematical and modeling origin, but it is also not an optimal paramenter for calibration. At first the thickness of the reaction layer changes with k_{max}, because it is inversely proportional to k_{max} (see Section 3.2.4). This can be a problem for small k_{max} values which lead to thick reaction layers.

6.5 Model Calibration

For a better understanding of the second trouble with large k_{max} values the following is worth mentioning: Simulations have shown that with regard to the finite elements for the value of k_{max} the following choice is reasonable

$$k_{max} = k_{global}/d_{elem}. \qquad (6.4)$$

Here, d_{elem} is the average diameter of the finite elements in the used mesh, and k_{global} is a constant value independent of the mesh fineness.

Due to the mesh dependence of k_{max} its variation is limited. In the experiments it was found out that the value of k_{max} can not be increased arbitrarily. For larger values than suggested in (6.4) the numerical formulation becomes instable. In contrast to this, small k_{max} values are not a problem. Therefore, k_{max} is not a suitable parameter for the model calibration of a potentially because of thick reaction layer (small value) and numerical instability (large value).

After excluding two of the three parameters, the last parameter which is the surface oxidant concentration C^{Sur} is investigated. On surfaces which have contact with the oxidizing atmosphere the oxidant concentration is used as a Dirichlet boundary condition. The key idea is to modify C^{Sur} in order to calibrate the oxide thickness of the simulated oxidation process over time for different oxidation conditions. From the physical aspect a higher surface oxidant concentration means that a larger number of oxidants diffuse to the Si/SiO$_2$-interface and react with silicon, which results in a faster oxidation rate.

6.5.2 Calibration Concept and Example

It was found with experiments that the best results are obtained if C^{Sur} consists of a constant part $C^\star C_A$ and an η-dependent part $C^\star C_B \eta^{pow}$ so that the effective surface concentration can be written as a function of η

$$C^{Sur} = C^\star (C_A + C_B \eta^{pow}). \qquad (6.5)$$

Here $C^\star$ is the standard oxidant concentration in the gas atmosphere as used in the Deal-Grove model. C_A, C_B, and pow are the calibration parameters. Because the value of η is changed during the oxidation process, the value of C^{Sur} is also changing with time. For the onward process η goes toward 0 on the surface and so the second term for C^{Sur} disappears.

As example for the above described calibration concept a (111) oriented and $0.4\,\mu$m height silicon block is wet oxidized and the oxide thickness over time for different temperatures is calibrated. The bottom surface is fixed, the lateral surfaces can only move vertically and on the upper surface a free mechanical boundary condition is applied. Only the upper surface of the body has contact with the oxidizing atmosphere. The oxide thickness is measured between the upper surface and the η-level of 0.5.

SIMULATION OF THERMAL OXIDATION WITH FEDOS

In the calibration process the values of the parameters C_A, C_B, and pow are determined with the help of the in-house tool SIESTA (Simulation Environment for Semiconductor Technology Analysis) [108], so that the thickness values of the simulated oxide layers agree with the calculated physical reference values up to approximately 500 nm at any time for a temperature range of 900–1100 °C. The temperature dependent diffusion coefficient $D(T)$ is calculated as explained in (3.13). The other two model parameter $k_{max} = 60$ s^{-1} and $C^* = 3 \cdot 10^7 \frac{\text{part}}{\mu m^3}$ are kept constant over the whole temperature range.

It was found that in case of wet oxidation the value of the parameter $pow = 0.16$ in (6.5) can be hold constant for the temperature range of T = 900–1100 °C. Furthermore, the experiments show that the parameter C_A can be brought to a linear and the parameter C_B can be brought to a parabolic dependence on temperature, which is described by

$$C_A(T) = -3.34 + 4.4 \cdot 10^{-3} T \quad \text{and} \tag{6.6}$$

$$C_B(T) = -2.15 + 1.67 \exp\left(22.77 \cdot 10^{-6}(T - 900°C)^2\right). \tag{6.7}$$

The expressions for C_A and C_B were found empirically. Their values over temperature are plotted in Fig. 6.11. C_A and C_B do not have a physical background, they are pure fitting parameters.

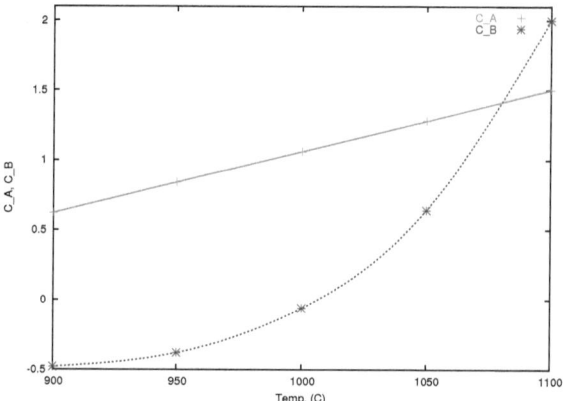

Figure 6.11: Parameters C_A and C_B over temperature.

In case of wet oxidation the three Figs. 6.12a–6.12c show vicegerent for all other temperatures that the formula for C^{Sur} with its parameters C_A, C_B, and pow leads to an excellent agreement between the calculated reference curves and the measured simulation curves. The oxide thickness values for the reference curves are calculated with the Deal-coefficients [51]. The coefficients C_A, C_B, and pow can be also found without problems for other process conditions (e.g. dry oxidation), and so the calibration with C^{Sur} always works well.

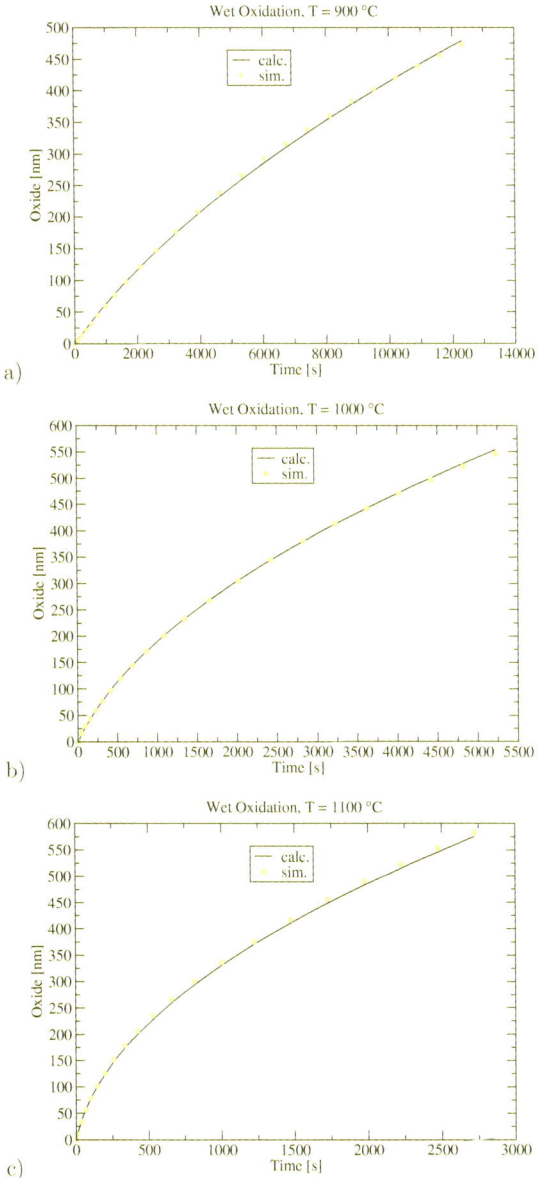

Figure 6.12: Oxide thicknesses over time at 900, 1000, and 1100 °C.

SIMULATION OF THERMAL OXIDATION WITH FEDOS

6.6 Comparison with a Two-Dimensional Simulation

The LOCOS structure shown in Fig. 6.3 is in principle a two-dimensional structure with a $0.4\,\mu$m stripped mask. Therefore, the three-dimensional simulation results from FEDOS can be compared with a two-dimensional oxidation simulation. For the stress dependent simulation with FEDOS a wet oxidation with a period of 20 minutes at $1000\,^\circ$C was assumed. The same parameters are used for an oxidation simulation on an equivalent two-dimensional structure with the commercial process simulation program DIOS [14]. The DIOS output is shown in Fig. 6.13. As illustrated in Fig. 6.10 the results from FEDOS are in good agreement with DIOS.

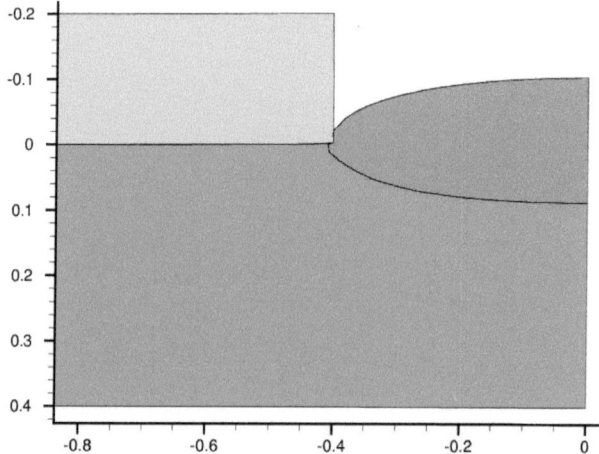

Figure 6.13: Two-dimensional oxidation simulation with DIOS.

Chapter 7

Stress Dependent Oxidation

STRESS is essential for thermal oxidation, because the oxidation process is considerably influenced by stress. Stress is always built up, if the volume increase of the new oxide is prevented somewhere from expanding as desired. During the oxidation process there can be a lot of stress sources like nitride masks, adjacent structures or oxidation of concave corners.

The oxidation process is stress dependent, because stress has an impact on the oxidant diffusion and the chemical reaction, which are both strongly reduced with stress and so the oxidation rate is also decreased in areas with compressive stress. For high stresses the oxidation process can be de facto even stopped. Thereby, for the simulation of stress dependent oxidation the model from Chapter 3 must be extended. In this chapter also the influence of stress is investigated with this extended model and the simulation results of representative examples are shown.

7.1 Oxidation Modeling with Stress

There are two parameters in the oxidation model, which are influenced by stress. The first one is the stress dependent diffusion coefficient [109, 110]

$$D(p,T) = D_0(T) \exp\left(-\frac{p V_D}{k_B T}\right). \tag{7.1}$$

Here $D_0(T)$ is the low stress diffusion coefficient (3.9), p is the pressure in the respective material, V_D is the activation volume, k_B is the Boltzmann's constant, and T is the temperature in Kelvin.

The second parameter is the stress dependent strength of a spatial sink

$$k(\eta, p) = \eta(\vec{x}, t) \, k_{max} \exp\left(-\frac{p V_k}{k_B T}\right). \tag{7.2}$$

STRESS DEPENDENT OXIDATION

Both parameters are exponentially reduced with pressure, which is only valid for values $p \geq 0$ [111].

With these two stress dependent parameters the three main equations in the oxidation model, which describe the oxidant diffusion (3.2), the η-dynamics (3.5), and the volume increase (3.8), become

$$D(p,T)\,\Delta C(\vec{x},t) = k(\eta,p)\,C(\vec{x},t), \tag{7.3}$$

$$\frac{\partial \eta(\vec{x},t)}{\partial t} = -\frac{1}{\lambda}k(\eta,p)\,C(\vec{x},t)/N_1, \quad \text{and} \tag{7.4}$$

$$V_{rel}^{add} = \frac{\lambda-1}{\lambda}\Delta t\, k(\eta,p)C(\vec{x},t)/N_1. \tag{7.5}$$

The stress is generally described with the formula

$$\tilde{\sigma} = \mathbf{D}(\tilde{\varepsilon} - \tilde{\varepsilon}_0) + \tilde{\sigma}_0, \tag{7.6}$$

where $\tilde{\varepsilon}_0$ stands for the desired volume increase

$$\varepsilon_{0,xx} = \varepsilon_{0,yy} = \varepsilon_{0,zz} = \tfrac{1}{3}V_{rel}^{add}, \tag{7.7}$$

and $\tilde{\varepsilon}$ represents the actual volume expansion, because ε_{ij} are the partial derivatives of the actual displacements (3.21). On a finite element the mechanical problem $\mathbf{K^e}\,\vec{d^e} = \vec{f^e}$ is loaded by the desired volume increase ($\varepsilon_{0,ii}$-values) which leads to the internal forces

$$\vec{f_{int}^e} = \mathbf{B^T}\mathbf{D}\,\tilde{\varepsilon}_0^{\,e}\,V^e. \tag{7.8}$$

The actual displacements $\vec{d^e}$ are obtained after solving the mechanical system (see Fig. 5.5). With these results the actual strains can be calculated

$$\tilde{\varepsilon}^e = \mathbf{B}\,\vec{d^e}, \tag{7.9}$$

and the stress on an element can be determined with (7.6).

A worth mentioning aspect is the visco-elastic stress computation in the FEDOS simulation procedure. For the actual time step n the visco-elastic stress $\tilde{\sigma}^n$ is the sum of a dilatation and a deviatoric part, because $\mathbf{D} = \mathbf{D}_{dil} + \mathbf{D}_{dev}$ as depicted in Section 3.2.5.2. Therefore, also the residual stress $\tilde{\sigma}_0^n$ for the actual time step n consists of a dilatation and a deviatoric part so that

$$\tilde{\sigma}_0^n = \tilde{\sigma}_{0,dil}^n + \tilde{\sigma}_{0,dev}^n. \tag{7.10}$$

The components of the actual residual stress tensor are build up from the $(n-1)$ previous time steps Δt according to (7.11) for the dilatation and (7.12) for the deviatoric part [112]

$$\sigma_{0,dil}^n = \sum_{i=1}^{n-1}\sigma_{dil}^i = \sigma_{dil}^{n-1} + \sigma_{0,dil}^{n-1}, \tag{7.11}$$

$$\sigma_{0,dev}^n = \sum_{i=1}^{n-1} \sigma_{dev}^i \exp\left(-\frac{(n-i)\cdot\Delta t}{\tau}\right) = \left(\sigma_{dev}^{n-1} + \sigma_{0,dev}^{n-1}\right)\exp\left(-\frac{\Delta t}{\tau}\right). \tag{7.12}$$

An important characteristic of visco-elastic materials is the stress relaxation of the deviatoric stress components over time with the Maxwellian relaxation time constant τ, as given in (7.12). The recursive form for residual stress calculation in the right hand side of (7.11) and (7.12) offers the benefit that the residual stress parts $\sigma_{0,dil}^n$ and $\sigma_{0,dev}^n$ at actual time step n can be simply computed by adding the components σ_{dil}^{n-1} and σ_{dev}^{n-1} from the last step $(n-1)$ to the already existing residual stress parts $\sigma_{0,dil}^{n-1}$ and $\sigma_{0,dev}^{n-1}$ determined at previous step $(n-1)$.

In contrast to stress the pressure needed for (7.1) and (7.2) is a scalar. It is positive, if the com-pressive stress components which have a negative sign, are predominant. So pressure always has an opposite sign compared to stress. The pressure is the average of the stress tensors trace

$$p = -\frac{\text{Trace}(\tilde{\sigma})}{3} = -\frac{\sigma_{xx} + \sigma_{yy} + \sigma_{zz}}{3}. \tag{7.13}$$

7.2 Stress Calculation Concept for Simulation

Normally only a small part of the whole real structure is investigated by simulation because of limited computer resources and desired short simulation times. Mostly the simulation of this small part delivers the needed information, because most structures have only few areas of interest or they are repeating. Therefore, also for the oxidation process the simulated domain is a three-dimensional cut of the complete structure.

Such a cut is shown in Fig. 7.1. This example represents a piece of the silicon substrate with (1.2×0.3) µm floor space where two thirds of the length are covered with a 0.15 µm thick silicon nitride mask. Only the upper surface has contact with the oxidizing ambient. The body has plain side walls which must not be deformed by simulation. This means that the four side walls are not allowed to move in their normal directions, as demonstrated in Fig. 7.2.

For the 125% additional volume of the newly formed oxide in (7.7) an isotropic expansion is assumed. This means that all strain components $\varepsilon_{0,ii}$ are equal. Because of the prevented movements of the simulation domain in the normal directions of the side walls the volume can not expand in the xy-plane, only in z-direction. The mechanical boundary conditions and the isotropic approach build up an enormous stress (pressure) in the whole oxide layer (see Fig. 7.3). In the mathematical formulation (7.6) this effect can be explained by the fact that $\varepsilon_{xx} = \varepsilon_{yy} = 0$.

The resulting high pressure all over the new generated oxide layer has the fatal effect on stress dependent simulation that the oxidation process is de facto stalled after a few

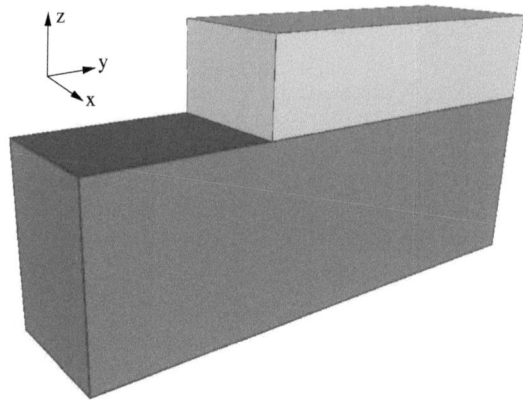

Figure 7.1: Structure with plain side walls for oxidation simulation.

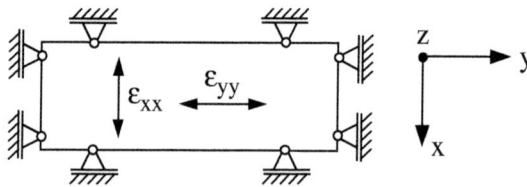

Figure 7.2: Side walls are restricted in movements to avoid their deformations.

time steps. The high pressure in the SiO_2-layer is in principle a wall for oxidant diffusion and chemical reaction, because both are decreased exponentially with pressure. Thereby, even for a long oxidation time the oxide thickness is minimal (see Fig. 7.3), and so the simulation results are totally wrong.

A possibility to solve this problem starts with the following considerations. For a plain surface (xy-plane) which is oxidized the oxide nearly grows stress-free only in the normal z-direction. In that case the isotropic approach for the volume increase is not correct, because it should be $\varepsilon_{0,xx} = \varepsilon_{0,yy} = 0$ and $\varepsilon_{0,zz} = V_{rel}^{add}$ in order to get the correct displacements of the new oxide in z-direction. For this purpose the isotropic approach should be modified. The question is how this can be performed automatically, because the displacements $\vec{d}$ are the results of the mechanical problem and the strains $\varepsilon_{0,ii}$ are the inputs. On the other side for the simulation of different structures the isotropic approach is the most general one.

It was found that the best strategy is to calculate the displacements in two steps. In the first step, denoted with $^{(1)}$, the displacements $\vec{d}^{e,(1)}$ on a finite element are calculated with

7.2 Stress Calculation Concept for Simulation

Figure 7.3: High pressure in the whole oxide layer due to isotropic expanding approach.

the universal isotropic approach

$$\varepsilon_{0,xx}^{e,(1)} = \varepsilon_{0,yy}^{e,(1)} = \varepsilon_{0,zz}^{e,(1)} = \tfrac{1}{3} V_{rel}^{add}, \tag{7.14}$$

The actual strains for the first step can be calculated after solving the mechanical problem with the results $\vec{d}^{e,(1)}$ by

$$\tilde{\varepsilon}^{e,(1)} = \mathbf{B}\,\vec{d}^{e,(1)}. \tag{7.15}$$

The idea now is to use these strains from the first step to load the mechanical problem for the second step. The actual strain components $\tilde{\varepsilon}_{ii}^{e,(1)}$ show in which directions the volume of a finite element can expand easily and in which ones it can extend hardly or is even blocked ($\tilde{\varepsilon}_{ii}^{e,(1)} = 0$). The actual expansion s_i in each direction x, y, and z can then be expressed by

$$s_i^{(1)} = \frac{\tilde{\varepsilon}_{ii}^{e,(1)}}{\tilde{\varepsilon}_{xx}^{e,(1)} + \tilde{\varepsilon}_{yy}^{e,(1)} + \tilde{\varepsilon}_{zz}^{e,(1)}} \qquad \text{with} \qquad i = x, y, z. \tag{7.16}$$

As example it is assumed that $s_x = 0$, $s_y = 0.2$, and $s_z = 0.8$. This means that the volume expansion is blocked in x- and prevented in y-direction. In z-direction there is the least resistance and so 80% of the actual volume increase happens there.

The minimal pressure in the elements can be reached, if the ratio of the input strains $\varepsilon_{0,ii}$ is the same as the percentage of the actual expansions s_i, because the ratio of the input strain components would be the same like the percentage of possible volume expansion in each direction. Therefore, the input strains for the second mechanical step are exactly weighted with the actual expansions from the first step in oder to get a minimal pressure

STRESS DEPENDENT OXIDATION

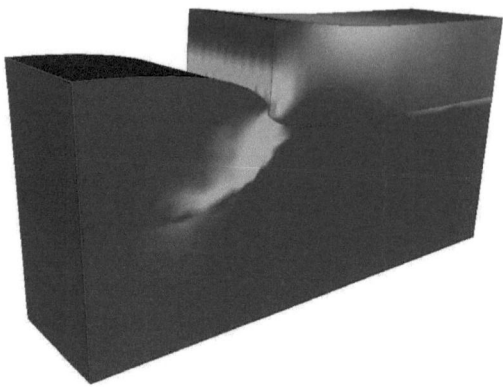

Figure 7.4: Pressure distribution with the two-step stress calculation concept.

in the increasing volume

$$\varepsilon_{0,ii}^{e,(2)} = s_i^{(1)} V_{rel}^{add} = \frac{\tilde{\varepsilon}_{ii}^{e,(1)}}{\tilde{\varepsilon}_{xx}^{e,(1)} + \tilde{\varepsilon}_{yy}^{e,(1)} + \tilde{\varepsilon}_{zz}^{e,(1)}} V_{rel}^{add}, \qquad \text{with} \qquad i = x, y, z. \qquad (7.17)$$

The strains $\tilde{\varepsilon}_0^{e,(2)}$ load the mechanical problem with (7.8) for the second step. After solving the mechanical system again with (7.15) the actual strains $\tilde{\varepsilon}^{e,(2)}$ and therefore the final stress (pressure) for each finite element can be found with the conventional stress formula

$$\tilde{\sigma}^e = \mathbf{D}(\tilde{\varepsilon}^{e,(2)} - \tilde{\varepsilon}_0^{e,(2)}) + \tilde{\sigma}_0^e. \qquad (7.18)$$

With the above described two-step stress calculation concept the pressure distribution in the simulated oxide domain meets the real physical conditions, as demonstrated in Fig. 7.4, because it avoids unnatural stresses which only come from the inappropriate modeling approach (isotropic expansion approach) and a simulation effect (cut structure where the side walls are not allowed to move in normal direction). Therefore, with this method and its right pressure distribution, the simulation with stress dependent parameters is treated properly, as displayed in Fig. 6.10.

7.3 Representative Examples

The advanced oxidation model, as described in Chapter 3, 5, and 7, is applied on two different structures to simulate the oxidation process. In addition to the results for the stress dependent simulation, as modeled in Section 7.1, also the results without stress dependent parameters as specified in Section 3.2.1 and 3.2.2 and the pressure distribution in the materials are of interest.

7.3 Representative Examples

7.3.1 First Example

The first three-dimensional example is an initial silicon block with $(0.8 \times 0.8)\mu m$ floor space which is covered with a 0.15 μm thick L-shaped Si_3N_4-mask as shown in Fig. 7.5. The Si_3N_4-mask prevents the oxidant diffusion on the subjacent silicon layer, because here only the upper surface has contact with the oxidizing ambient.

The plain side walls of the body must not be deformed by simulation. Therefore the mechanical boundary conditions are set such that the four side walls are not allowed to move in their normal directions. The bottom surface is fixed and on the upper surface a free mechanical boundary condition is applied.

The simulation result of the oxidation process after a time t_1, which is the η-distribution, is shown in Fig. 7.6. Here blue is pure SiO_2 ($\eta = 0$), red is the pure silicon substrate ($\eta = 1$), and at the Si/SiO_2-interface there is the reaction layer with a spatial finite width ($0 < \eta < 1$) as explained in Section 3.1. Due to the L-shaped mask the effect of the three-dimensional oxidation process is pronounced, because the shape of the SiO_2-region and the deformations are not continuous in any direction.

For a more physical interpretation of the simulation results with a sharp interface between silicon and SiO_2 the two regions are extracted from the η-distribution by determining that $\eta \leq 0.5$ is SiO_2 and $\eta > 0.5$ is silicon as shown in Fig. 7.7. For an optimal comparison of the geometry before and after oxidation as well as the influence of stress, Figs. 7.5–7.10 have the same perspectives and the same proportions.

7.3.2 Stress Dependence

In order to demonstrate the importance of the stress dependence the results with and without the impact of stress are compared. Since the oxidant diffusion and the chemical reaction are exponentially reduced with the hydrostatic pressure in the material, the oxidation process itself is highly stress dependent.

Fig. 7.8 shows the pressure distribution in the materials, where the positive pressure regions are displayed in red (more red means more pressure). It can be seen that the highest pressure in SiO_2 is under the edge of the Si_3N_4-mask, because in this area the stiffness of the mask prevents the desired volume expansion of the newly formed SiO_2. Due to the mentioned stress dependence the oxidation rate in these areas is considerably reduced (see Fig. 7.7). The stiffness of the Si_3N_4-mask is approximately six times larger than the stiffness of SiO_2 and therefore the displacements in SiO_2 are also much more larger than in the Si_3N_4-mask which leads to the well known bird's beak effect.

If the stress dependence is not taken into account for the simulation of the oxidation process, the simulation results do not agree with the real physical behavior, because the oxide region is too large. In this case the oxidant diffusion and the chemical reaction

STRESS DEPENDENT OXIDATION

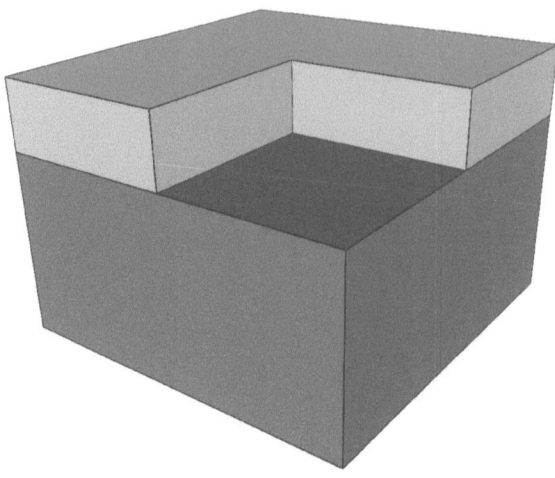

Figure 7.5: Initial structure of the Si-Si$_3$N$_4$-body before thermal oxidation.

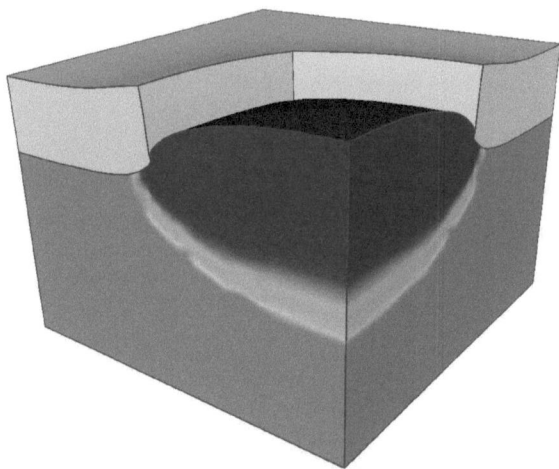

Figure 7.6: η-distribution and reaction layer after thermal oxidation at time t_1.

also occur under the Si$_3$N$_4$-mask without restriction and therefore the SiO$_2$-region at the same oxidation conditions is much more expanded than with the stress dependence as demonstrated in Fig. 7.9. In addition, the larger forces under the Si$_3$N$_4$-mask, which result

7.3 Representative Examples

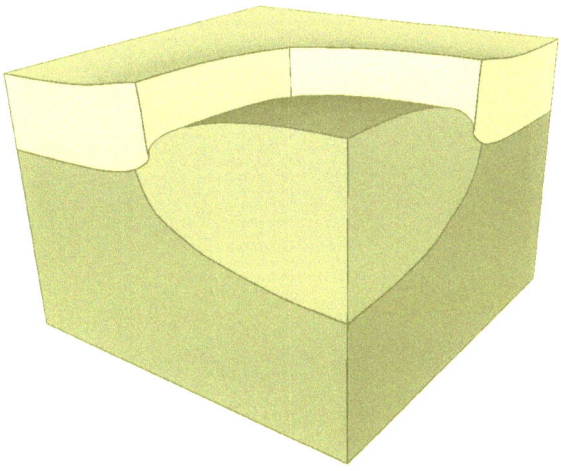

Figure 7.7: SiO$_2$-region (sharp interface) with stress dependent oxidation at time t_1.

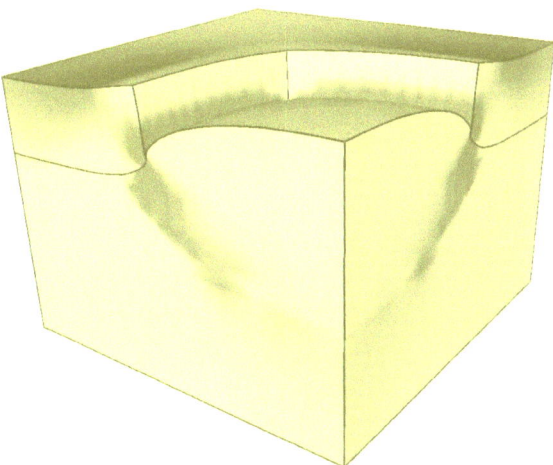

Figure 7.8: Pressure distribution with stress dependent oxidation at time t_1.

from the larger pressure domain in this area (see Fig. 7.10), cause larger displacements of the mask.

STRESS DEPENDENT OXIDATION

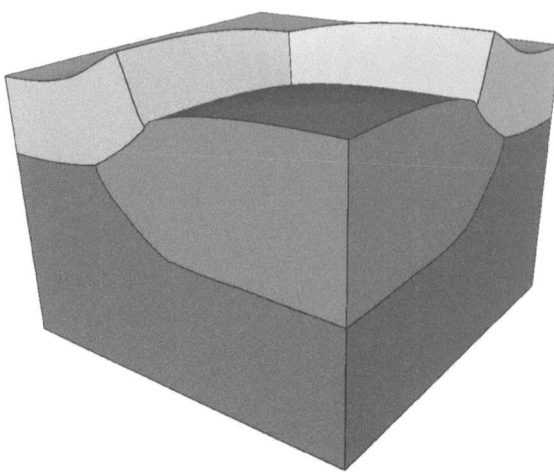

Figure 7.9: SiO$_2$-region (sharp interface) without stress dependent oxidation at time t_1.

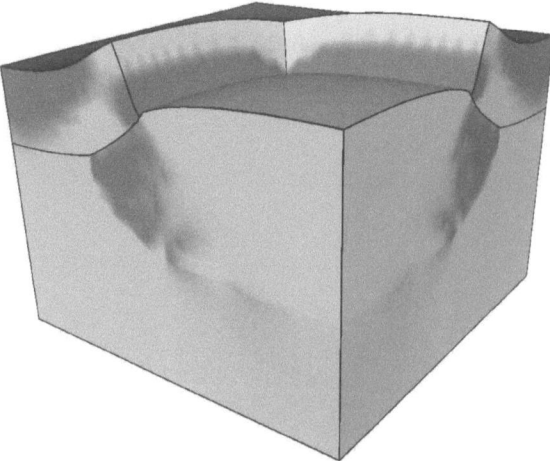

Figure 7.10: Pressure distribution without stress dependent oxidation at time t_1.

7.3.3 Second Example

The second example is a commercially fabricated EEPROM-cell where thermal oxidation is an inherently important step in the production procedure. Fig. 7.11 is the top view of a SEM picture which shows a field of 6 EEPROM-cells a few process steps after oxidation. In this picture A is the active area of the cell which is crossed by the unshaped floating gate C. The rest of the cell area is the field oxide B made by the oxidation process. The swells of the active area are provided for the contacts. In Fig. 7.11 the active area A is surrounded by a light line. This line marks the original area of the already removed Si_3N_4-mask. Because of the bird's beak effect during the oxidation step the active area is smaller than the masked one.

The oxidation process is only simulated on a cut of the whole field of EEPROM-cells, because the structure is repeating. The analyzed structure which is part of a cell, is marked with the rectangle in Fig. 7.11 and has $(1.5 \times 1.0)\mu m$ floor space. This complex structure is displayed on Fig. 7.12 before oxidation, where the upper layer is a 0.15 μm thick Si_3N_4-mask. The area which is masked with Si_3N_4 is not oxidized, and results in the active area A of the EEPROM-cell after removal of the mask (see Fig. 7.11).

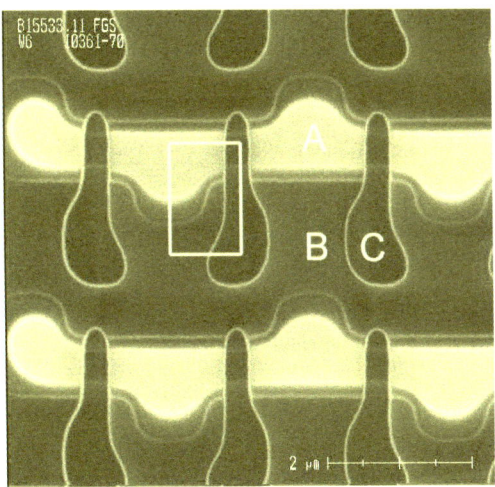

Figure 7.11: SEM picture (top view) of the EEPROM-cell field.

Fig. 7.13 shows the results of the simulated oxidation process on the EEPROM-structure after a time t_a. For an optimal illustration of the stress dependence effect in the simulation results a 45°-cut in the area of the convex mask curve (see Fig. 7.12) is performed. With this three-dimensional example it is demonstrated again that the stress dependence must

STRESS DEPENDENT OXIDATION

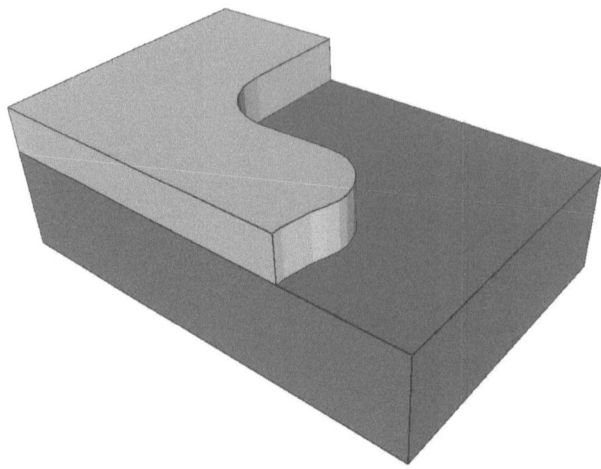

Figure 7.12: Initial structure of the analyzed structure before thermal oxidation.

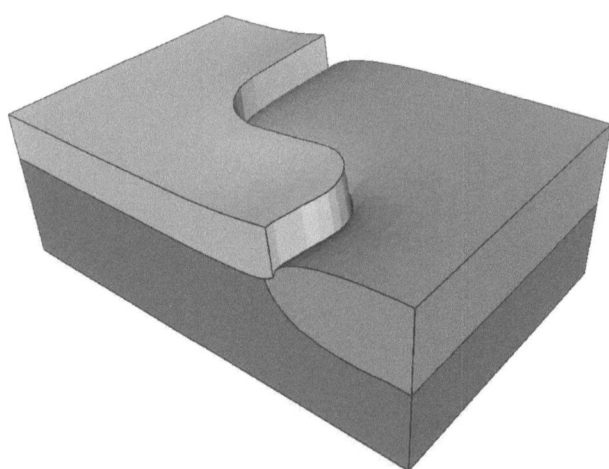

Figure 7.13: SiO$_2$-region with stress dependent oxidation at time t_a.

be taken into account in the oxidation model (see Fig. 7.14). Otherwise the oxidation rate is not reduced in the pressure domains, which leads to wrong simulation results because of the too large oxide region and deformations (see Fig. 7.16).

7.3 Representative Examples

Figure 7.14: SiO$_2$-region and deformation with stress dependent oxidation at time t$_a$.

Figure 7.15: Pressure distribution with stress dependent oxidation at time t$_a$.

STRESS DEPENDENT OXIDATION

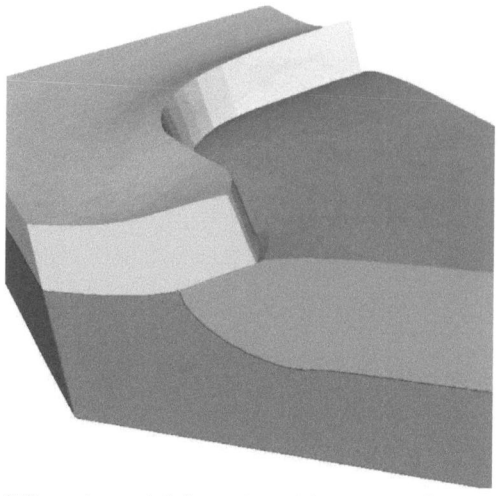

Figure 7.16: SiO$_2$-region and deformation without stress dependent oxidation at time t$_a$.

Figure 7.17: Pressure distribution without stress dependent oxidation at time t$_a$.

Chapter 8

Thermo-Mechanical Stress in Interconnect Layouts

THERMO-MECHANICAL STRESS plays an important role in the development of electromigration failure mechanisms in interconnect layouts. Electromigration is the main reliability factor in integrated circuits, because it can cause a system failure after some undetermined operating time [113]. Electromigration is the mass transport of a metal due to the momentum transfer between conducting electrons and diffusing metal atoms [114]. Electromigration in principle exists, whenever a current flows through a metal wire, because the electrons which flow through a metal wire collide with metal atoms. This collisions produce a force on the metal atoms in the direction of the electron flow (for n-type materials and opposite for p-type materials).

Electromigration is only significant at high current densities (e.g. 10^6 A/cm^2) in metals [115]. The magnitude of its force is proportional to the current density [116]. Because of its material transport, electromigration leads to void formation and void growth where material is depleted [117]. The void causes a large increase in the electric resistance [118], even up to values that the connection practically fails. The void can also reach so large dimensions that the interconnect is broken [119]. Opposite, in points with material accumulation a cracking of the dielectric and a formation of an extrusion can occur which results in a short between adjacent lines.

In advanced semiconductor manufacturing processes, copper has replaced aluminum as the interconnect material of choice. Despite its greater fragility in the fabrication process [120], copper is preferred for its superior conductivity. It is also intrinsically less susceptible to electromigration [121], but electromigration is still an everpresent challenge for device fabrication. Since copper diffuses into silicon and most dielectrics, copper lines must be encapsulated with metallic (like TaN or TiN) and dielectric (such as SiN or SiC) diffusion barriers in order to prevent corrosion and electrical leakage between adjacent copper leads. Because of the different adjacent materials with its different thermal expansion coefficients, thermo-mechanical stresses are preassigned.

Besides the current density and high temperature, latter one caused by Joule self-heating, thermo-mechanical stress is one of the important electromigration promoting factors [115, 122]. For the accurate simulation of the electromigration reliability, the influence of mechanical stress should be taken into account, but state of the art simulators lack this capability. The simulation should predict the time to failure and should locate possible critical points in an interconnect structure. Critical points are locations where electromigration promoting factors like current density, temperature and thermo-mechanical stress have high values. Furthermore, interconnect simulation also includes the prediction of void nucleation, void evolution and resistance change. All these above listed problems about electromigration and its modeling are already described in Chapter 4 in [123]. Therefore, this chapter is only focused on the simulation of thermo-mechanical stress.

8.1 Simulation Procedure

Thermo-mechanical simulation demands a temperature distribution in the structure. It was found out that the electrical characteristics of the complete system do not considerably change with stress during standard operation. Therefore, the simulation can be separated into an electro-thermal and a thermo-mechanical part within small time periods as long as there is no void nucleation in the interconnect lines or the passivation is not broken.

In the simulation sequence as displayed in Fig. 8.1, the first part is the three-dimensional transient electro-thermal simulation of the interconnect structure in order to calculate the temperature distribution. Additionally this simulation delivers the potential and the current density. With the temperature distribution from the first part, the three-dimensional thermo-mechanical simulation can be performed subsequently. With the electro-thermal and the thermo-mechanical simulation all necessary capabilities for the rigorous simulation of electromigration are available.

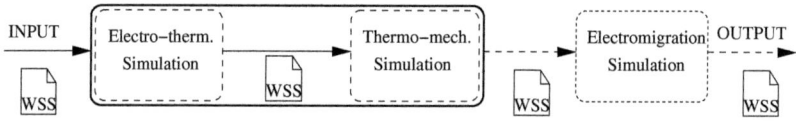

Figure 8.1: Simulation sequence and data flow.

For the electro-thermal investigation the simulator STAP [124] from the *Smart Analysis Programs* package, which has also been developed at the Institute for Microelectronics, is used. The other simulations can be carried out by the diffusion and oxidation simulator

FEDOS. For the straightforward data exchange between the different simulators, WSS files are used, because STAP as well as FEDOS can handle the WSS format.

Although STAP is also based on FEM, it is not appropriate to use it for thermo-mechanical simulations, because STAP is specialized and optimized for fast and accurate electro-thermal simulations. An extension of STAP to a more universal tool which can also handle mechanical problems would reduce its performance significantly. So even with the necessary data exchange the decoupled simulations with STAP and FEDOS are more efficient than a coupled simulation only performed by STAP.

8.1.1 Electro-Thermal Simulation

The electro-thermal simulation is performed with the simulator STAP which uses also the finite element method for the calculation of the electric potential and temperature distribution. For the numerical calculation of Joule self-heating effects, caused by the electric current flow through the wire, two partial differential equations have to be solved [125, 126]. The first one describes the electric subproblem

$$\operatorname{div}(\gamma_E \operatorname{grad} \varphi) = 0. \tag{8.1}$$

The electric potential φ needs to be solved only inside domains composed of electrically conducting material (γ_E represents the electrical conductivity). On the surface of the conductors three types of boundary conditions are allowed:

- Dirichlet - a constant potential is specified
- Neumann - vanishing current density is specified
- Floating potential - the total current is specified and the potential is forced to be the same all over the boundary area.

The next step is to compute the power loss density p_D described by

$$p_D = \gamma_E \, (\operatorname{grad} \varphi)^2. \tag{8.2}$$

In addition the heat conduction equation has to be solved in order to obtain the distribution of the temperature T in the whole interconnect structure

$$\operatorname{div}(\gamma_T \operatorname{grad} T) = c_p \rho_m \frac{\partial T}{\partial t} - p_D, \tag{8.3}$$

where γ_T represents the thermal conductivity, c_p the specific heat, and ρ_m the mass density.

The temperature dependence of the thermal and electrical conductivities is modeled with second order approximations:

$$\gamma(T) = \gamma_0 \frac{1}{1 + \alpha(T - T_0) + \beta(T - T_0)^2}. \tag{8.4}$$

In (8.4) γ_0 is the electrical or thermal conductivity at a temperature T_0 of 300 K, α and β are the linear and quadratic temperature coefficients. This makes the problem non-linear. Since the non-linearity is relatively weak, a simple iterative relaxation method is used which quickly converges to the solution, usually after 3-6 iterations.

8.1.2 Thermo-Mechanical Stress Simulation

High tensile stresses in the copper interconnects can cause break-up of the material and development of voids [127]. On the other side compressive stresses can induce the generation of extrusions. In case of temperature changes thermo-mechanical stress is build up because of (significant) different thermal expansion coefficients of adjacent materials.

The thermo-mechanical stress simulation is carried out with the program package FEDOS. The modeling of thermo-mechanical stress is similar to the stress calculation during thermal oxidation as described in Section 7.1. For the assumed elastic materials the stress tensor can be written in the form

$$\tilde{\sigma} = \mathbf{D}(\tilde{\varepsilon} - \tilde{\varepsilon}_0) + \tilde{\sigma}_0. \tag{8.5}$$

The main components of the residual stress tensor $\tilde{\varepsilon}_0$ include the desired volume change of the material caused by the temperature change. The strain components $\varepsilon_{0,ii}$ (i stands for x, y, and z) are linearly proportional to the temperature T in the material

$$\varepsilon_{0,ii} = \alpha_m \left(T - T_0 \right), \tag{8.6}$$

where α_m is the thermal expansion coefficient for the respective material and T_0 is the ambient temperature assumed with 300 K.

The temperature change loads the mechanical problem $\mathbf{K}^e \, \vec{d^e} = \vec{f^e}$ on every finite element, because the internal force vector is

$$\vec{f^e_{int}} = \mathbf{B^T} \mathbf{D} \, \tilde{\varepsilon}_0^{\,e} \, V^e. \tag{8.7}$$

The actual displacements $\vec{d^e}$ are obtained after solving the linear global equation system for the mechanics. With these results the actual strains can be calculated with $\tilde{\varepsilon}^e = \mathbf{B} \, \vec{d^e}$ and so the stress on an element can be determined with (8.5).

With the stress in principle also the hydrostatic pressure is given with the formula

$$p = -\frac{\mathrm{Trace}(\tilde{\sigma})}{3} = -\frac{\sigma_{xx} + \sigma_{yy} + \sigma_{zz}}{3}. \tag{8.8}$$

8.2 Demonstrative Example

A three-dimensional interconnect layout with (3.0×4.2) μm floor space, as displayed in Fig. 8.2, is investigated by the two previously defined models. In this structure the bottom layer material is silicon (Si). Above the silicon layer there is a silicon dioxide (SiO_2) layer, where two copper (Cu) lines are embedded. Between the copper lines and the silicon dioxide is a very thin titanium nitride (TiN) passiviation layer. This passiviation layer prevents the diffusion of copper into the silicon dioxide during the manufacturing process [128].

As shown in Fig. 8.2, the silicon dioxide layer and the copper lines are covered by a silicon nitride (Si_3N_4) layer which separates the next upper located SiO_2 layer from the lower one. In the second upper SiO_2 layer three copper lines are embedded. These three copper lines are transverse compared to the two subjacent copper lines. This upper SiO_2 layer is also covered with silicon nitride. On the top of the layout is a third SiO_2 layer.

In Fig. 8.3 a cut through the interconnect structure given in Fig. 8.2 is presented. As evident from Fig. 8.3 an upper transverse copper line is connected with a lower copper line by a so-called via. The other two copper lines shown in this figure are connected in the same way. The third transverse upper copper line (see Fig. 8.2) does not have an interconnection to another line.

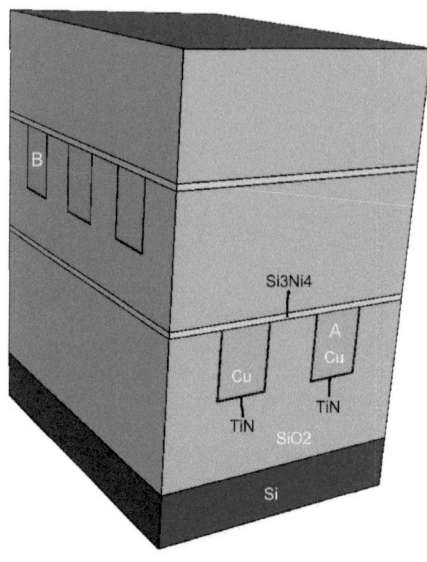

Figure 8.2: Investigated complete interconnect structure.

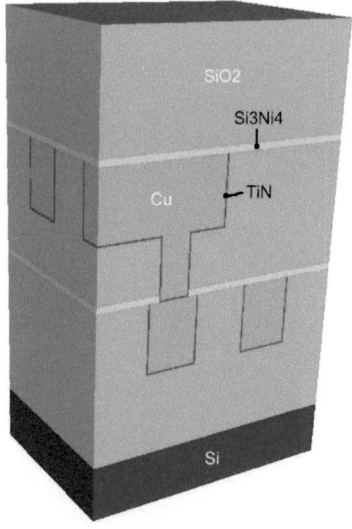

Figure 8.3: Cut through the interconnect layout.

8.2.1 Simulation Results

From the simulation aspect the temperature and stress distribution in the given interconnect structure at two different points of time are of interest. For the simulation a potential difference of 7 mV between point A and B in the first interconnect, as marked in Fig. 8.2, is assumed. The other interconnects are assumed to be inactive.

8.2.1.1 Temperature Distribution

If STAP with its electro-thermal model is applied to this interconnect structure, the obtained output is the temperature distribution in the structure. In this analysis it is assumed that the bottom of the silicon layer is connected with a cooling element which holds the temperature at 320 K. For the simulation the electric and thermal conductivities given in Table 8.1 are used. Because copper is a metal and an excellent conductor, it has the best thermal and electric conductivity regarding feasible materials for interconnect metals.

Table 8.1: Electric and thermal conductivities at 300 K [129].

	Cu	Si	SiO_2	Si_3N_4	TiN
γ_E [S/m]	5.26×10^7	0.0	0.0	0.0	1.66×10^5
γ_T [W/mK]	400.0	1.35	1.39	12.07	48.25

Due to the Joule self-heating in wires with a current flow, the hottest regions are around the active copper line. This is the reason, why after a time of 30 μs the highest temperature (351 K) is in the inner layers which are surrounded by the active copper line, as shown in Fig. 8.4. The relatively high thermal conductivity of copper causes that the temperature values in the copper lines are rather uniform, and so they are not included in Fig. 8.4. In Fig. 8.5 it is demonstrated that after a longer operating time of 100 μs the self-heating has increased the temperature to 376 K.

In Fig. 8.7 the maximal temperature versus operating time in the interconnect structure is plotted. It can be seen that after approximately 450 μs the self heating effect reaches a stable steady state. Fig. 8.6 shows the temperature distribution in the steady state where the maximal temperature has reached a steady value of 396 K.

8.2.1.2 Pressure Distribution

For the pressure calculation, the influence of the temperature on the mechanical parameters can be neglected, because the temperatures are so low that they do not change the material condition perceivably [130]. Also the influence of pressure on the electric and thermal parameters can be neglected for such pressure values [131]. Therefore, a coupling

THERMO-MECHANICAL STRESS IN INTERCONNECT LAYOUTS

of the electro-thermal and thermo-mechanical system is not necessary and the decoupled alternating solving of the temperature and pressure distribution is acceptable here. With the obtained temperature distribution the mechanical problem can be set up as described in Section 8.1.2. As applied mechanical boundary conditions the bottom surface is fixed and the other surfaces are free. For the simulation the Young modulus E, Poisson ratio ν, and the thermal expansion factor α_m given in Table 8.2 are used.

Table 8.2: Mechanical parameters

	Cu	Si	SiO$_2$	Si$_3$N$_4$	TiN
E [GPa]	115	180	73	380	600
ν [-]	0.34	0.22	0.17	0.27	0.25
α_m [10^{-6}/K]	17.7	2.7	0.55	3.3	9.4

The distribution of pressure is an important quantity for electromigration, because failure risks are increasing with larger pressure. Therefore, the copper lines and their vias are the most interesting regions. The thermal expansion coefficient of copper is the largest and it is enormous compared with the main embedding material silicon dioxide (see Table 8.2). This larger coefficient of the copper lines demands more volume expansion than the other surrounding materials in the heated structure. This means that the copper lines with their vias can not expand as desired and compressive stress is built up.

Fig. 8.8 shows the simulation results of the pressure distribution in the copper lines and their vias at time 30 μs. It can be seen that the via is a high pressure region. The first reason is that the via has less chance to expand in vertical direction because of the over- and underlying copper lines which also have the same demand to extend. The other explanation is the confinement of the via by the passivation layer (see Fig. 8.3) which was made of titanium nitride (TiN). The passivation layer is thin, but the stiffness (Young modulus) of TiN is more than five times larger than the copper one and so this layer is able to prevent the volume increase. The proof is that the largest pressure (188 MPa) develops at the bottom of the vias, because it is confined with the passivation layer. The pressure in the bottom region is larger than on the top, where is no limiting TiN-layer.

After a longer operating time of 100 μs the higher temperature in the structure (see Fig. 8.5), causes that the maximum pressure in the via is increased to 354 MPa, as displayed in Fig. 8.9. The pressure distribution in the copper lines and vias is nearly the same as at time 30 μs. As illustrated in Fig. 8.10 the pressure reaches a maximum of 479 MPa in the steady state.

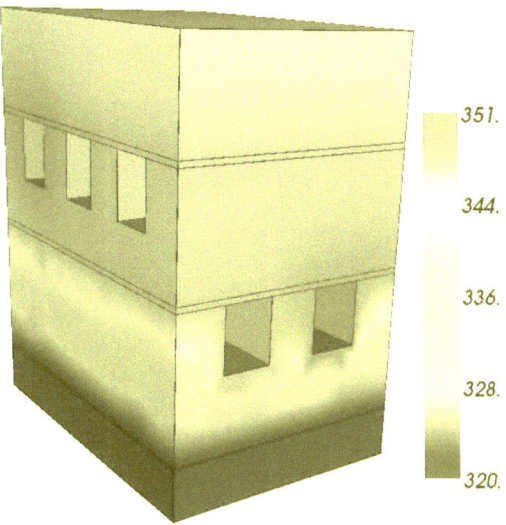

Figure 8.4: Temperature distribution in the silicon dioxide (SiO$_2$) and silicon nitride (Si$_3$N$_4$) layers in Kelvin [K] at time 30 µs.

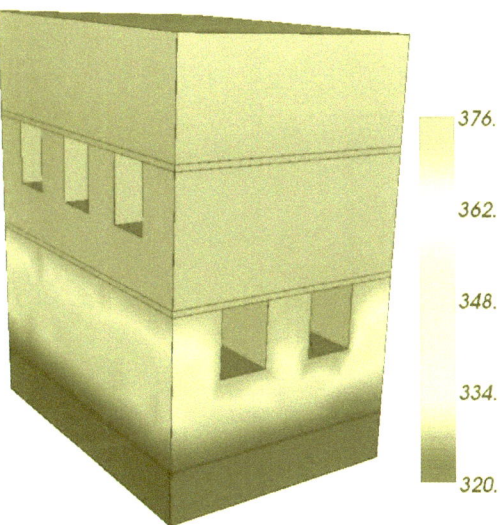

Figure 8.5: Temperature distribution in the silicon dioxide (SiO$_2$) and silicon nitride (Si$_3$N$_4$) layers in Kelvin [K] at time 100 µs.

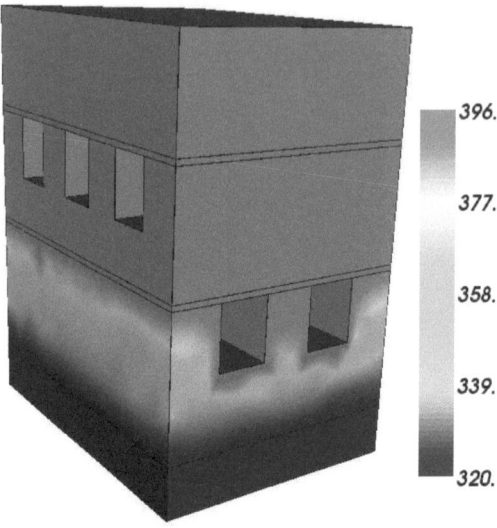

Figure 8.6: Temperature distribution in the silicon dioxide (SiO$_2$) and silicon nitride (Si$_3$N$_4$) layers in Kelvin [K] in the steady state.

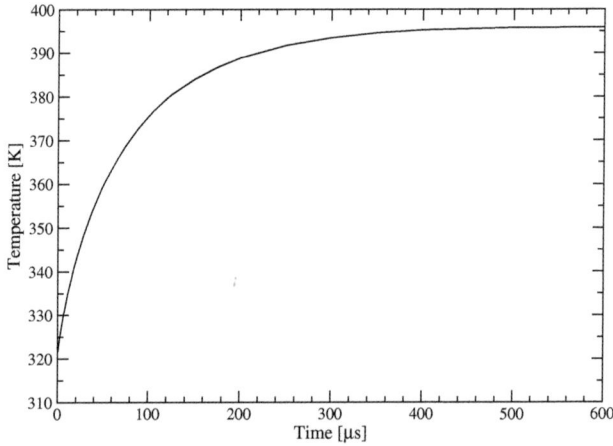

Figure 8.7: Maximal temperature versus time in the interconnect structure.

8.2 Demonstrative Example

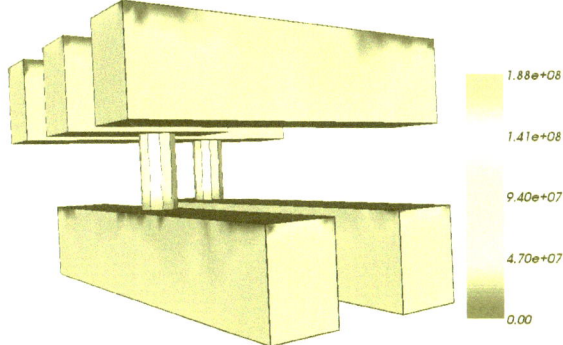

Figure 8.8: Pressure distribution in the copper lines and their vias in Pascal [Pa] at time 30 μs.

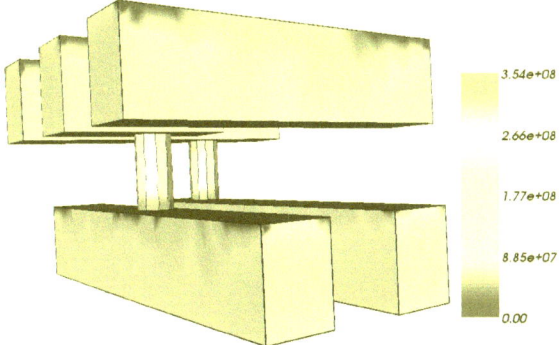

Figure 8.9: Pressure distribution in the copper lines and their vias in Pascal [Pa] at time 100 μs.

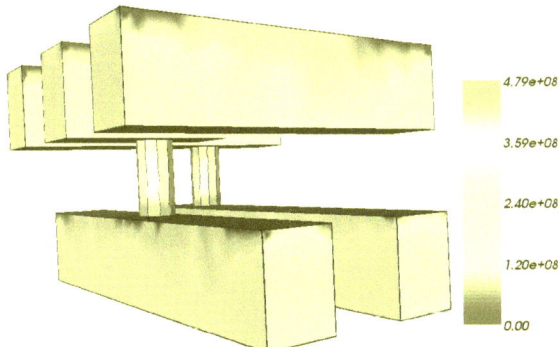

Figure 8.10: Pressure distribution in the copper lines and their vias in Pascal [Pa] in the steady state.

Chapter 9

Intrinsic Stress Effects in Deposited Thin Films

THIN FILM DEPOSITION is a widely used technique for the fabrication of MEMS (Micro-Electro-Mechanical Systems). This technique is required to manufacture free-standing structures which can induce or sense a mechanical movement. During the deposition process of thin layers and aftermath an intrinsic stress is generated. In subsequent process steps, after removal of the underlying sacrificial layer, the (stressed) deposited layer which is an important component of the desired MEMS device, is left free-standing. As a consequence the process induced stress can relax and deform the deposited layer in an undesirable way.

Polycrystalline silicon-germanium (poly-SiGe) has been promoted as an attractive material suitable as structural layer for several MEMS applications [132]. Poly-SiGe is a good alternative to polycrystalline silicon (poly-Si), because it has similar properties. The same good mechanical and electrical properties can be obtained with poly-SiGe at much lower temperatures (down to 400 °C) compared to poly-Si (above 800 °C). These low processing temperatures enable MEMS post-processing on top of MOS without introducing significant changes in the existing MOS fabrication processes. The sacrificial layer is normally made of silicon dioxide (SiO_2), because this material can then be etched with a high selectivity towards the structural layer by the use of hydrogen fluoride (HF).

Different aspects of the connection between microstructure and stress have been investigated in the past 30 years. The focus was mostly on some specific grain-grain boundary configurations in early or mature stages of microstructure evolution [133]. As a result there exist numerous models derived on the basis of continuum mechanics, which are applicable only for highly simplified situations. On the other side a group of researchers, mostly mathematicians, has developed complex models for describing morphology of the microstructural evolution, a development which culminates in multi-level set models of

INTRINSIC STRESS EFFECTS IN DEPOSITED THIN FILMS

grain evolution [134, 135]. These models can reproduce the realistic grain boundary network in a high degree, but they do not include stress [135]. The goal of this work is the integration of microstructure models which describe strain development due to grain dynamics in a macroscopic mechanical formulation. This strain loads the mechanical problem which provides a distribution of the mechanical stress and enables the calculation of displacements in the MEMS structure.

9.1 Cantilever Deflection Problem

An everlasting challenge for MEMS engineering is to fabricate a free-standing cantilever without any unwanted deflection, but in practice the thin films can not be deposited stress free. These effects of stress in free-standing MEMS structures can be demonstrated most plausibly with the cantilever deflection problem. Such a deflection of a 400 μm long and 10 μm thick fabricated cantilever is shown in Fig. 9.1.

Figure 9.1: Cantilever deflection. Courtesy of IMEC/Gregory van Barel.

9.1.1 Principle of Cantilever Deflection

Fig. 9.2 shows the schematic structure of a free-standing cantilever, where the SiGe structural layer is deposited on the SiO_2 sacrificial layer. In this case it is assumed that there does not exist any stress gradient in the SiGe film, and so no deformation of the released cantilever occurs after removal of the underlying sacrificial material by etching.

Normally the structural layer is not deposited stress free and therefore the bending of the released cantilever depends on the stress distribution over the layer thickness before

9.1 Cantilever Deflection Problem

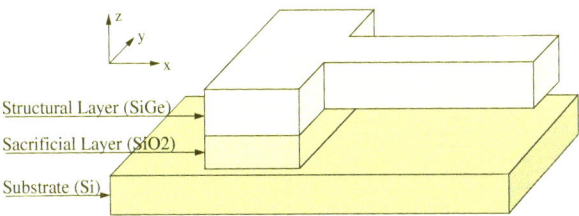

Figure 9.2: Structure of a free-standing cantilever without stress gradient.

release. In this context only the stress above and under the neutral bending line is responsible for the direction of the deflection, because stress causes forces which result in the bending moments regarding the neutral bending line. If the sum of the moments under the neutral bending line is larger than the above one, than the deflection is upward, otherwise downward. The neutral bending line is always located in the middle of the cantilever and, therefore, its location is changing with the layer thickness.

For an assumed linear stress gradient Γ over thickness and a rectangular cross-section area A of the beam with a width w and thickness t, the deflection of the cantilever $\delta(x)$ at position x is

$$\delta(x) = \frac{Mx^2}{2EI} = \frac{\Gamma}{2E}x^2, \tag{9.1}$$

where E is the Young modulus, $M = \Gamma\frac{wt^3}{12}$ is the bending moment and $I = \frac{wt^3}{12}$ is the moment of inertia. In (9.1) it can be seen that for a constant Γ the deflection at the end of the cantilever ($x=l$) increases quadratically with the length l and so $\delta(l) = \frac{\Gamma}{2E}l^2$.

Fig. 9.3 shows three possible forms of stress distribution and gradients in the fixed cantilever on the left and the corresponding direction of deflection on the right.

In the first situation with a positive stress gradient (Fig. 9.3a), the part of the beam above the neutral bending line is in a tensile state while the part below is in a dominant compressive state. Compressive stress in a body is an indication that the material has desire to expand, but the expansion is prevented. In the conventional declaration compressive stress always has a negative sign, so that the pressure can be positive as quoted in (7.14). Therefore, after release of the beam the upper tensile part can contract and the compressive one can dilate and, therefore, it is comprehensible that the deflection can only go upward.

The next case in Fig. 9.3b demonstrates that is is not necessary to have stresses with opposite sign on the two sides of the neutral bending line for a deflection. Here all stresses have compressive character, but due to the positive stress gradient the compressive stress values and the moments below the neutral bending line are larger than above. Therefore, there is also a deflection in the upward direction. In contrast to the previous configurations the stress gradient in the last case (see Fig. 9.3c) is negative. This situation is inverse to

INTRINSIC STRESS EFFECTS IN DEPOSITED THIN FILMS

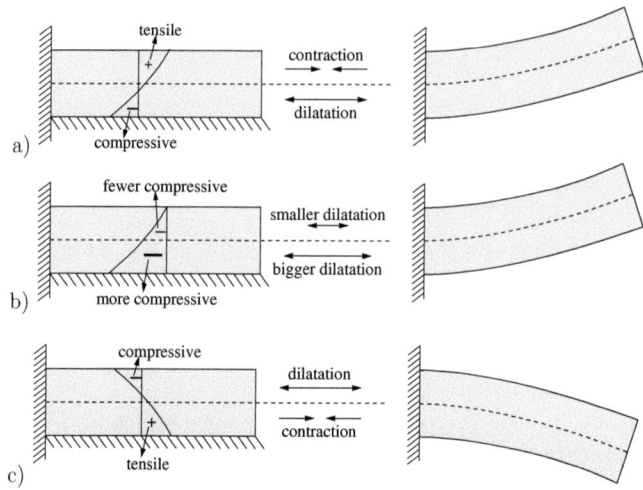

Figure 9.3: Various stress distributions and gradients in the fixed cantilever and their respective deflections after release.

the first one, because there is an upper compressive part and a bottom tensile part. In consequence the cantilever deflection goes downward after release.

9.1.2 Stress Distribution and Relaxation

As example for a positive stress gradient in thin films (see Fig. 9.3a), where only tensile stress was assumed, the stress distributions for a 1 mm long and 10 μm thick cantilever structure were simulated. As long as the cantilever is attached with the underlying SiO_2-layer, the deposited SiGe film is under stress and the cantilever can not deform. Because of the positive stress gradient the highest tensile stress, marked with red color, is on the top of the fixed cantilever as demonstrated in Fig. 9.4a. After removal of the sacrificial SiO_2-layer by etching, the SiGe beam is free standing. Now the cantilever can deform and the stress is relaxed, as shown in Fig. 9.4b. The intrinsic stress in the deposited SiGe layer is the driving force for the cantilever deflection. As listed in the next Section 9.2, there are a number of intrinsic stress sources.

9.2 Sources of Intrinsic Stress

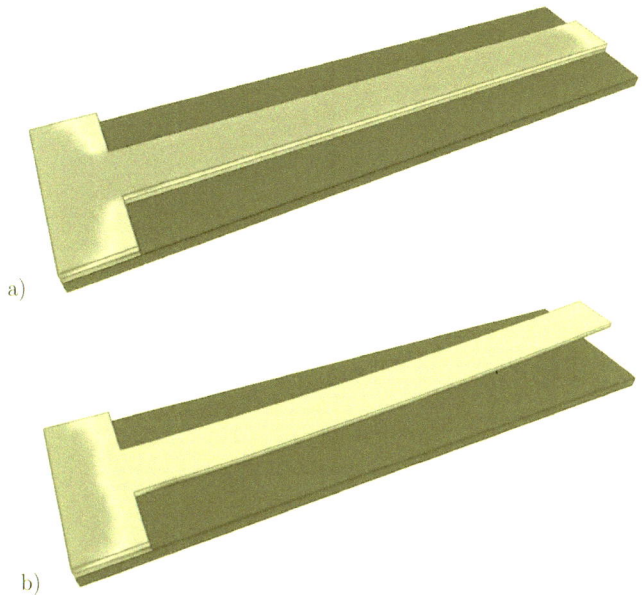

Figure 9.4: Stress distribution for the fixed a) and released b) 1 mm long cantilever. High stress areas are marked with red color.

9.2 Sources of Intrinsic Stress

In the first phase of the SiGe deposition process, islands with varying crystal orientation are formed and grow isotropically. These individual islands which first form on a substrate usually exhibit compressive stress [136]. In the course of further deposition these islands start to coalescence, which forces the islands to grow in the height instead of in a direction parallel to the substrate surface. The islands are subsequently transformed from an island shape to a grain-like shape. The orientation of the crystal structure in a single grain (e.g. perpendicular to the substrate surface) is independent of the neighboring grains, since due to the amorphous substrate, it is not possible to evolve a perfect crystal structure in the first atom layers [137].

For the stress aspect the deposition process plays a key role. At first it should be noted that the deposition takes place at elevated temperatures. When the temperature is decreased, the volumes of the grains shrink and the stresses in the material increase. Furthermore, the stress gradient and the average stress in the SiGe film depend on the Si-Ge ratio which can be controlled by the silane (SiH_4) and germane (GeH_4) flow, the substrate temperature, and the deposition technique which is usually LPCVD (low pressure

INTRINSIC STRESS EFFECTS IN DEPOSITED THIN FILMS

chemical vapor deposition) or PECVD (plasma enhanced chemical vapor deposition). It was observed that the average stress becomes more compressive, if the Ge concentration decreases [138]. Thus it is expected that a film with higher Ge concentration has a higher degree of crystallinity and larger grains, which leads to higher film density and to higher tensile stress.

The intrinsic stress observed in thin films has generally the following main sources [133]:

- Coalescence of Grain Boundaries:
  In the early stage of the film growth the individual grain islands grow, until they make contact to adjacent islands (see Fig. 9.5a). The isolated islands have a relatively high surface energy γ_s compared to the relatively low energy γ_i between the island interfaces. Therefore, the net free energy in the system can be reduced by replacing the surfaces by interfaces. If the gaps between the islands are small enough, cohesion begins to develop between the islands, and the system can lower its net free energy by closing up these gaps as depicted in Fig. 9.5b. In the course of zipping up the interfaces, the participating islands become elastically strained and a tensile stress is generated [139].

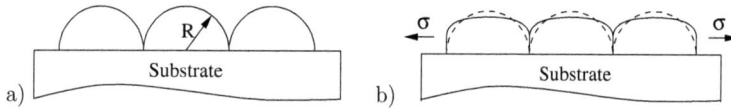

Figure 9.5: Principle of coalescence. Structure of the grain island before a) and after b) coalescence.

- Misfit Stress:
  The lattice constants for the thin film a_s and the substrate a_f are generally different (see Fig. 9.6a). Because of the deposition process the crystal lattice of the thin film and the substrate are forced to line up perfectly at the interface and stress arises as shown in Fig. 9.6b. The influence of these misfit stresses is only significant in the initial phase of thin film deposition [140], because of the local lattice adaption at the interface area. Furthermore, misfit stress can arise between the grain boundaries because of a different crystal orientation of neighboring grains.

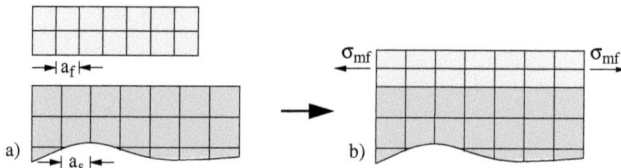

Figure 9.6: Different lattice constants a) leads to misfit stress in the film b).

- Annealing of the Film:
  An annealing step after deposition of metal films produces a better crystalline arrangement and an increase of the material density, which results in a shrinkage of the film [141]. As long as the film is attached to the substrate the film is prevented to shrink and a tensile stress is developed.

- Grain Growth:
  Due to the elimination of grain boundaries a minimum in the total energy of the system can be reached. So grain growth means that the volumes of the individual grains become larger and the number of grains and its boundaries decrease. The grain growth stops at this minimum energy. Since grain boundaries are less dense than the grain lattice [142], the elimination of grain boundaries leads to a densification of the film and, therefore, to a build up of tensile stress.

- Annihilation of Excess Vacancies:
  The annihilation and the dynamics of the crystal vacancies produce a local volume change which leads to stresses in the film when it is attached to the substrate. The vacancies annihilate in the grains, at the grain boundaries, at the free surface of the film, and at the surface of the internal cavities. If vacancies are annihilated at the free surface and at internal cavities, no stress is produced. When vacancies annihilate at a grain boundary, there is a gap. In addition vacancy annihilation in the grains leads to removal of atoms from the grain boundaries to the interior of the grains, which also leads to a gap. Both cases cause a motion of the crystals towards each other in order to close the gap. This would produce a planar contraction of the film, if it is not attached to the substrate. But since the substrate prevents contraction, a tensile stress is built up instead [133].

- Thermal Stress:
  This stress is caused by the different thermal expansion coefficients of the thin film and the substrate in case of a temperature change after deposition and the fact that at least a part of the film's base area is attached with the substrate. Therefore, thermal stress develops during cooling down to room temperature.

- Insertion of Excess Atoms:
  It is assumend that the film growth process can add atoms to the film in two ways [143]. Most of the material is added on the top surface by traditional crystal growth mechanisms, where each layer of atoms is deposited onto the underlying crystalline lattice. The second mechanism is the incorporation of excess atoms into the grain boundaries, which creates a compressive stress in the film [144].

9.3 Modeling of the Stress Sources

The goal of the modeling is to express the previous described microscopic stress phenomena with a macroscopic stress formulation for the respective source.

- *Coalescence of Grain Boundaries:*

The force which is generated by grain island impingement is [145]

$$f = \frac{4\pi}{3} R\gamma, \tag{9.2}$$

where R is the radius of the hemispherical island (see Fig. 9.5a) and γ is the surface energy of the contacting spheres.

The volume of material included in a hemispherical island upon coalescence is $\frac{2}{3}\pi R^3$. If the material in each island would spread uniformly over a $2R \times 2R$ square area on the substrate, the uniform depth h_c would be $h_c = \frac{1}{6}\pi R$. The area of a lateral face of this equivalent material block is $2R \times h_c$, so that the uniform tensile stress on this surface which produces the same resultant force as in (9.2) is $f/2Rh_c$ or $4\gamma/R$. Therefore, the average intrinsic stress caused by coalescence in the film with thickness h_c is [145]

$$\sigma_{xx}^{in} = \sigma_{yy}^{in} = 4\frac{\gamma}{R} = \frac{2\pi\gamma}{3h_c}, \qquad \sigma_{zz}^{in} = 0. \tag{9.3}$$

- *Misfit Stress:*

Misfit stresses occur in crystalline films because of the lattice mismatch at the interface between film and substrate. If only the film lattice would adjust to the substrate lattice as demonstrated in Fig. 9.6b, the misfit strain in the film would be $\varepsilon_{mf} = \frac{a_s - a_f}{a_f}$. But in reality the lattice of the film as well as the substrate are both adapted at the interface, which is characterized by the misfit parameter [146]

$$m = 2\frac{a_f - a_s}{a_f + a_s}, \tag{9.4}$$

where a_f and a_s are the lattice constants of the film and substrate (see Fig. 9.6a), respectively. The nonzero components of the misfit stress tensor are [140]

$$\sigma_{xx}^{in} = \sigma_{yy}^{in} = \frac{Em}{1-\nu^2}, \qquad \sigma_{zz}^{in} = \nu\sigma_{xx}^{in}. \tag{9.5}$$

- *Grain Growth:*

During grain growth some grain boundaries and their volumes disappear. Assume that V_0 is the pure crystal volume, where the excess volume of the grain boundaries is not included, and L_1 is the average grain diameter. Then the grain boundary area per unit volume is $6/L_1$ for spherical grains. If Δa is the excess volume per unit area of the grain boundary, the total excess volume for the grain boundaries in a volume V_0 is [133]

$$V^{xs} = V_0 \frac{6}{L_1} \Delta a. \tag{9.6}$$

9.3 Modeling of the Stress Sources

The total grain volume is

$$V_t = V_0 + V^{xs} = V_0\left(1 + \frac{6\Delta a}{L_1}\right). \tag{9.7}$$

The normalized volume change due to the disappearance of grain boundaries would be

$$\Delta V_n^* = \frac{V_t - V_0}{V_0} = \frac{6\Delta a}{L_1}. \tag{9.8}$$

If the grain grows to a new diameter L_2, the normalized volume change is [133]

$$\Delta V_n = 6\Delta a\left(\frac{1}{L_2} - \frac{1}{L_1}\right). \tag{9.9}$$

As defined in (3.37) the strain components ε_{xx}, ε_{yy}, and ε_{zz} are one third of the normalized volume change $\varepsilon_{xx} = \varepsilon_{yy} = \varepsilon_{zz} = \frac{\Delta V_n}{3}$. Therefore the intrinsic tensile stress associated with the grain growth is [133]

$$\sigma_{xx}^{\text{in}} = \sigma_{yy}^{\text{in}} = \sigma_{zz}^{\text{in}} = \frac{2E}{1-\nu}\left(\frac{1}{L_2} - \frac{1}{L_1}\right)\Delta a. \tag{9.10}$$

- *Excess Vacancy Annihilation:*

The gaps at the grain boundaries are closed by stretching the grains. The stress calculation is in principle the same as for the grain growth. If Ω_v is the vacancy volume and Ω_a is the atomic volume, then for a number of ΔC vacancies which annihilate per unit volume, the normalized volume change $\Delta V_n = \Delta C(\Omega_a - \Omega_v)$. Therefore, the intrinsic stress caused by vacancy annihilation is given by [133]

$$\sigma_{xx}^{\text{in}} = \sigma_{yy}^{\text{in}} = \sigma_{zz}^{\text{in}} = \frac{E}{1-\nu}\frac{\Delta C(\Omega_a - \Omega_v)}{3}. \tag{9.11}$$

Since during vacancy annihilation the vacancies diffuse to grain boundaries, the intrinsic stress can also be described with the more detailed and diffusion affiliated formulation [133]

$$\sigma_{xx}^{\text{in}} = \sigma_{yy}^{\text{in}} = \sigma_{zz}^{\text{in}} = \frac{4E\Omega_a}{L(1-\nu)}\sqrt{\frac{D_V t}{\pi}}(C_i - C_{\text{gb}}). \tag{9.12}$$

Here D_V is the vacancy diffusivity within the grain and L is the grain diameter. C_i is the vacancy concentration inside the grain and C_{gb} in the grain boundary, respectively.

- *Thermal Stress:*

The developed intrinsic stress due to thermal mismatch in the film and in the substrate material is

$$\sigma_{xx}^{\text{in}} = \sigma_{yy}^{\text{in}} = \sigma_{zz}^{\text{in}} = B\alpha(T - T_0), \tag{9.13}$$

where $B = (3\lambda + 2\mu)/3$ is the bulk modulus with the Lamé constants λ and μ, α is the thermal expansion coefficient, and T_0 is the ambient temperature.

INTRINSIC STRESS EFFECTS IN DEPOSITED THIN FILMS

For a uniform film the thermal stress is also uniform over the thickness. As in case of cantilevers, where after release still a part of the beam length is clamped at the bottom (see Fig. 9.2), this uniform stress has also a bending effect.

- *Insertion of Excess Atoms:*

The insertion of excess atoms into the grain boundaries creates a compressive stress in the film [144]

$$\sigma_{xx}^{in} = \sigma_{yy}^{in} = \sigma_{zz}^{in} = \frac{E}{1-\nu} \frac{\Delta C_i \Omega_e}{3}, \qquad (9.14)$$

where Ω_e is the volume of an excess atom and ΔC_i is the number of excess atoms which are inserted per unit volume.

In this work a methodology to predict a qualitative strain or stress curve over the film thickness was found. This methodology is based on the knowledge of the different intrinsic stress sources, the observed deflection characteristic of the deposited thin film material, and human mind. In the state of the art of development this methodology can not weight the influence of the different single stress sources in the different film thicknesses automatically. In order to find the strain or stress curve automatically a more advanced thin film stress model is necessary. The development of a thin film model, which takes all stress sources into account and weights them for different film materials and process conditions automatically, should be done in future work.

9.4 Investigation of Fabricated Cantilevers

The main purpose of cantilever simulation is to predict the deflection for different geometries (e.g. length and thickness), mechanical boundary conditions, and deposition process parameters. In the following, fabricated cantilevers with a cross section as described in Section 9.4.1 are investigated.

9.4.1 Cross Section

The cross section of the investigated cantilever structures is shown in Fig. 9.7. At the lower part of this SEM picture one can see the silicon substrate with a 250 nm thick sacrificial SiO_2-layer on it. Above the SiO_2-layer the picture shows the bottom part of the deposited poly-SiGe film. This multilayer film has a germanium concentration x between 62 and 65% in the layers. The Young modulus for silicon germanium $E_{SiGe} = E_{Si}(1-x) + E_{Ge}\, x$ varies only between 146 and 148 GPa under the assumption that E_{Si} is 173 GPa and E_{Ge} is 132 GPa, respectively.

The multilayer SiGe film is deposited in three steps:

1) At first a PECVD seedlayer with 95 nm thickness is deposited as nucleation layer for

9.4 Investigation of Fabricated Cantilevers

the following LPCVD layer, because the nucleation on the substrate with LPCVD needs much more time.

2) Then a 370 nm thick LPCVD layer is deposited in order to help crystallizing the top PECVD layer. Crystalline material has much more the desired properties than an amorphous one.

3) In the last step a PECVD layer with the desired film thickness (for example 10 μm) is deposited. PECVD films grow very fast, namely at 120–130 nm/min, while LPCVD films have only a deposition rate between 16–19 nm/min [138].

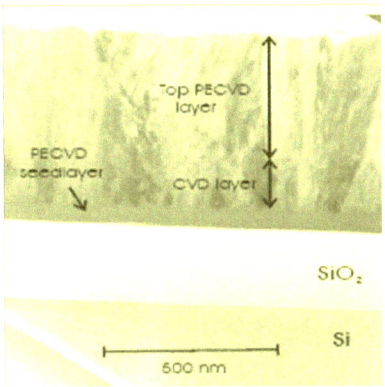

Figure 9.7: Cross section of the poly-SiGe multilayer. Courtesy of IMEC/Gregory van Barel.

9.4.2 Strain Curve

The developed methodology to treat thin film stress is applied to the experimental setting presented in [147]. In this experiment a 10 μm thick SiGe film was deposited on a SiO_2 sacrificial layer, as described above. After removal of this sacrificial layer, the deflection of the free 1 mm long cantilever was measured at different thicknesses from 10 down to 1 μm. The smaller thicknesses were made by thinning. It was observed that the deflection increases exponentially with reduced thickness.

The intrinsic strain curve for this SiGe multilayer film (see Fig. 9.8), which is qualitatively predicted by the found methodology, was calibrated according to the measurement results. Since the SiO_2 layer is amorphous, no misfit stress can arise here. It is worth mentioning that intrinsic compressive strain which loads a mechanical problem, must have a positive sign, because compressive materials want to expand. Compressive strain has only the same negative sign as stress, if a material is compressed by external forces.

INTRINSIC STRESS EFFECTS IN DEPOSITED THIN FILMS

The highest intrinsic compressive strain value with 0.08 is at the bottom of the SiGe film. This can be explained with a compressive stress exhibition of the individual islands which first form on the sacrificial layer [136], and with the insertion of excess atoms. Thermal stress can also be compressive. Within the next 800 nm of the film the strain plunges down to a minimum of 1.3×10^{-3} because of the tensile stress source in the deposited material, namely the coalescence of grain boundaries, the grain growth, and the excess vacancy annihilation. In the rest of the film there is a slow increase of the compressive part. For this phenomenon it is assumed that the grains tend to grow isotropically, but due to their neighbors they are prevented to extend in the plane and they are forced to grow into the height instead, which leads to compressive stress.

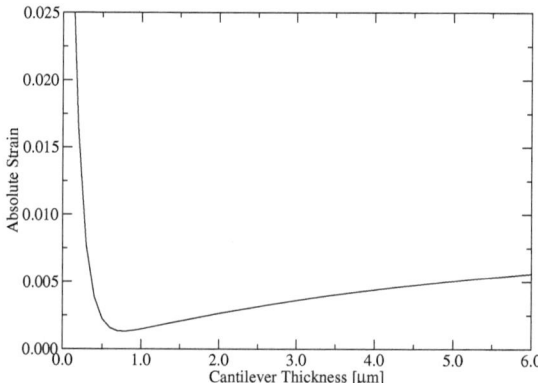

Figure 9.8: Strain versus thickness in the SiGe multilayer thin film.

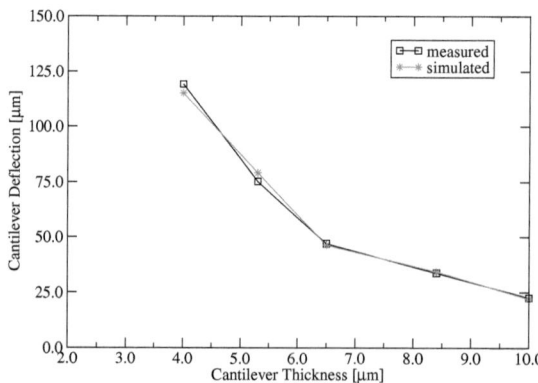

Figure 9.9: Measured and simulated cantilever deflections for different thicknesses.

The large compressive strain at the bottom of the SiGe film explains the very large deflections for thin cantilevers. At first the neutral bending line which is located midway, is moving with the cantilever thickness, and secondly the stiffness is decreased for thinner cantilevers. This strain curve was used to simulate the deflections for various thicknesses for the 1 mm long cantilever structure as shown in Fig. 9.4. As demonstrated in Fig. 9.9, the simulated cantilever deflections show good agreement with the experimentally determined deflections.

9.4.3 Practical Example

As practical example for the simulation procedure a fabricated cantilever as shown in Fig. 9.10 is used. In this SEM picture which shows an array of unreleased cantilevers with different lengths, the surrounded SiO_2 is already removed so that the side walls of the cantilevers lie free. The etching process was stopped before the sacrificial layer is removed and, therefore, the SiGe cantilevers are still fixed. The light material which separates and frames the cantilevers is also SiGe with the same composition as for the cantilevers. In Fig. 9.10 the selected structure is marked with a yellow rectangle. This cantilever is 900 μm long, 50 μm wide, and 6 μm thick. The multilayer cross section of this SiGe cantilever is the same as displayed in Fig. 9.7 and described in Section 9.4.1.

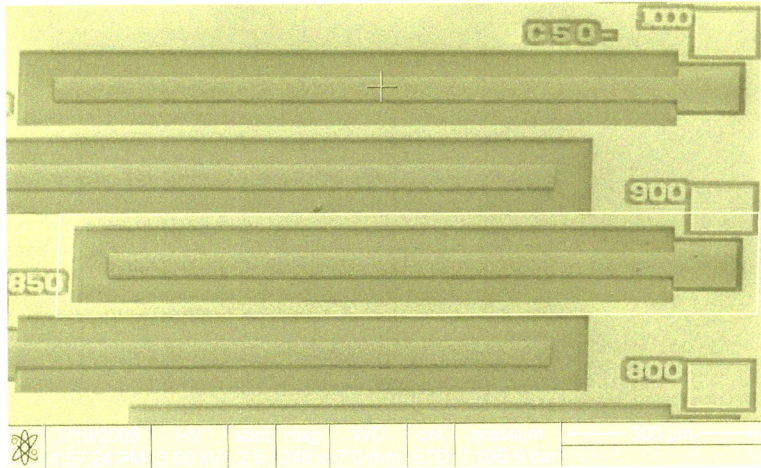

Figure 9.10: Array of unreleased cantilevers. Courtesy of IMEC/Gregory van Barel.

Fig. 9.11 shows the initial structure for the simulation with FEDOS, where the silicon substrate is green, the SiGe frame is blue and the cantilever is red. The dimensions of the simulated geometry are identical with the yellow framed structure in Fig. 9.10. The structure has a floor space of $(1120 \times 220)\,\mu\text{m}$. The strain curve (see Fig. 9.8) loads the deflection problem. The simulated deflection at the end of the 900 μm long and 6 μm thick cantilever is 44.6 μm. The structure after simulation with the deflected cantilever is displayed in Fig. 9.12. A cut of this deflected cantilever structure is shown in Fig. 9.13.

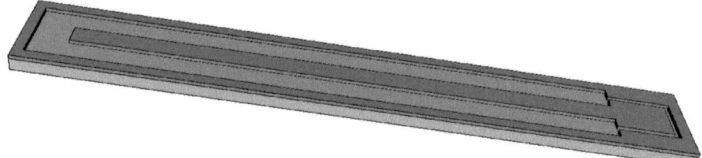

Figure 9.11: Initial structure with unreleased cantilever.

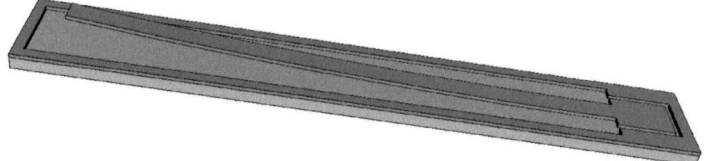

Figure 9.12: Cantilever structure after simulated deflection.

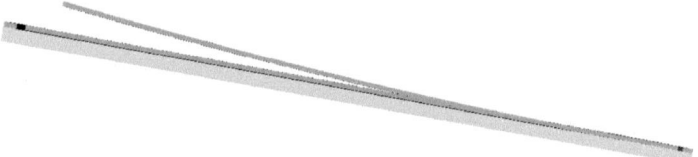

Figure 9.13: Cut of the deflected cantilever structure.

Chapter 10

Summary and Conclusions

THE THERMAL OXIDATION PROCESS was described, modeled, and simulated. The two main isolation techniques for neighboring MOS transistors were described as introduction. The differences in the process flow and final oxide shape of LOCOS and STI were demonstrated. The main feature of the LOCOS process is the bird's beak effect with its oxide field encroachment.

During the oxidation process a chemical reaction converts silicon into SiO_2 and the nearly sharp Si/SiO_2-interface moves from the surface into the silicon substrate. The formed SiO_2 has more than twice of the original volume of silicon, which is the main source of stress and displacements in the materials. The oxidation process depends on four parameters: the used oxidant species, the temperature and the pressure in the furnace, and the crystal orientation of the silicon substrate. Since the oxide growth rate is strongly temperature dependent for all species, in practice the oxide growth is mainly controlled only by the temperature.

An enhanced three-dimensional oxidation model was developed which is based on a diffuse interface with a reaction layer. This model takes into account that during oxidation the oxidant diffusion, the chemical reaction, and the volume increase occur simultaneously. The diffuse interface concept avoids the drawbacks of the moving boundary problem, complicated mesh algorithms, and enormous data update. Therefore, the enhanced model enables the simulation of even complex structures with a moderate demand on computer resources. Since SiO_2 and Si_3N_4 show visco-elastic behavior, besides an elastic also a visco-elastic formulation with a so-called effective shear modulus was introduced for the mechanics.

The effects of thermal oxidation of doped silicon material were described. Because of the built-up Si/SiO_2-interface segregation leads also to a redistribution of the dopands. For modeling this redistribution the five-stream diffusion model from Dunham was introduced.

SUMMARY AND CONCLUSIONS

In order to solve the mathematical formulation numerically, the finite element method was applied. The finite element discretization with tetrahedrons for the oxidant diffusion, the η-dynamics, and the mechanics was explained in detail. The principle of the assembling in order to built up a complete equation system and the handling of Dirichlet boundary conditions and mechanical interfaces was described.

The enhanced oxidation model was implemented in the in-house process simulation tool FEDOS. The architecture and main components of FEDOS were depicted and the simulation procedure for oxidation was explained. The mesh plays a key role for simulation, because the number of finite elements is always a compromise between accuracy and simulation time. This means that an acceptable accuracy should be reached with as small as possible number of elements. The most effective strategy found is to use a static mesh. It was demonstrated that in critical regions, e.g. along the edge of a mask, the mesh should be finer than in the rest of the structure. Due to the diffuse interface concept of the enhanced oxidation model, the procedure for a physical interpretation of the displayed simulation results with a sharp Si/SiO_2-interface was described. For practical applications of the oxidation model, the simple but effective model calibration with the surface oxidant concentration was shown.

Stress has a significant influence on the oxidant diffusion and the chemical reaction, and so also on the resulting oxide growth rate. A universal stress calculation concept for the simulation of stress dependent oxidation, where the stress in the structure is determined in two steps, was presented. The enhanced oxidation model was applied to simulate two three-dimensional (fabricated) structures. It was demonstrated that only when the stress dependence of the oxidation process is taken into account, the simulation results agree with the real physical behavior.

In copper interconnects stress is an important promoting factor for electromigration, which can lead to void formation and to failure of the interconnect. The procedure for the simulation of thermal stress in a representative interconnect structure was described. First, a electro-thermal simulation was performed in order to obtain the temperature distribution in the interconnect layout due to Joule self-heating. With this temperature distribution the thermal stress can be simulated. The reason for thermal stress are the different thermal expansion coefficients of the respective materials in the adjacent layers. The highest stress values in the interconnect structure were predicted at the bottom of the vias. Therefore, this is the most critical region for electromigration.

The effects of intrinsic stress in deposited thin films were discussed. A negative effect of stress in free-standing MEMS structures was demonstrated with the unwanted deflection of cantilever. For a linear stress gradient the deflection of the cantilever increases quadratically with the length. A number of sources which can generate tensile or compressive stress in the film were described. The whole intrinsic stress comes from microscopic effects like grain dynamics. Macroscopic mechanical formulations for the different intrinsic stress sources, which describe the stress development due to the microscopic effects, were listed. A methodology which can predict a qualitative strain or stress curve over the film thickness was developed. This methodology was applied to determine qualitatively the strain curve for a deposited multilayer SiGe film. The found strain curve was calibrated and applied to simulate the stress and deflection in a fabricated cantilever structure.

Bibliography

[1] A. Hössinger, *Simulation of Ion Implantation for ULSI Technology*. Dissertation, Institute for Microelectronics, Vienna University of Technology, 2000. http://www.iue.tuwien.ac.at/phd/hoessinger/.

[2] M. Radi, *Three-Dimensional Simulation of Thermal Oxidation*. Dissertation, Institute for Microelectronics, Vienna University of Technology, 1998. http://www.iue.tuwien.ac.at/phd/radi/.

[3] Stanford University, *SUPREM-IV*. http://www-tcad.stanford.edu/tcad/programs/suprem45.html.

[4] Stanford University, *TCAD*. http://www-tcad.stanford.edu/index.html.

[5] Integrated Circuits Laboratory, Standford University, *SUPREM-IV.GS, Two-Dimensional Process Simulation for Silicon and Gallium Arsenide*, 1993.

[6] Synopsys Inc., *SUPREM-IV, Two-Dimensional Process Simulation Program*, 2003.

[7] Wikipedia, *Semiconductor Process Simulation*. http://en.wikipedia.org/wiki/Semiconductor_process_simulation.

[8] Silvaco International, *ATHENA User's Manual, 2D Process Simulation Software*, 2004.

[9] Silvaco International, *Homepage*. http://www.silvaco.com.

[10] Wikipedia, *Silvaco*. http://en.wikipedia.org/wiki/Silvaco.

[11] Silvaco International, *ATHENA*. http://www.silvaco.com/products/process_simulation/athena.html.

[12] Synopsys Inc., *Taurus Process & Device, User Manual*, 2003.

BIBLIOGRAPHY

[13] R. Minixhofer, *Integrating Technology Simulation into the Semiconductor Manufacturing Environment*. Dissertation, Institute for Microelectronics, Vienna University of Technology, 2006. http://www.iue.tuwien.ac.at/phd/minixhofer/.

[14] Integrated Systems Engineering AG, *DIOS, ISE TCAD Release 10.0*, 2004.

[15] University of Florida, *FLOOPS Manual*. http://www.swamp.tec.ufl.edu/~flooxs/FLOOXS Manual/Intro.html.

[16] S. Cea and M. Law, "Three Dimensional Nonlinear Viscoelastic Oxidation Modeling," in *Proc. Int. Conference on the Simulation of Semiconductor Processes and Devices (SISPAD)*, pp. 97–98, 1996.

[17] Private communication with Prof. Mark Law in January 2007.

[18] Integrated Systems Engineering AG, *ISE News*, December 2003.

[19] Integrated Systems Engineering AG, *FLOOPS-ISE, ISE TCAD Release 10.0*, 2004.

[20] Synopsys Inc., *Homepage*. http://www.synopsys.com.

[21] Synopsys Inc., *TCAD Products*. http://www.synopsys.com/products/tcad/tcad.html.

[22] Synopsys Inc., *Newsletter*, December 2004. http://www.synopsys.com/products/tcad/pdfs/news_dec04.pdf.

[23] Synopsys Inc., *Sentaurus: Advanced Simulator for Process Technologies*. http://www.synopsys.com/products/tcad/pdfs/sprocess_ds.pdf.

[24] Synopsys Inc., *Newsletter*, October 2005. http://www.synopsys.com/products/tcad/pdfs/news_oct05.pdf.

[25] J. D. Plummer, M. D. Deal, and P. B. Griffin, *Silicon VLSI Technology: Fundamentals, Practice and Modeling*. New Jersey: Prentice Hall, 2000.

[26] T. Hori, *Gate Dielectrics and MOS ULSIs: Principle, Technologies and Applications*, vol. 34 of *Electronics and Photonics*. Berlin: Springer, 1997.

[27] B. El-Kareh, *Fundamentals of Semiconductor Processing Technologies*. Norwell: Kluwer Academic Publishers, 1995.

[28] C. R. Helms, "The Atomic and Electronic Structure of the Si-SiO$_2$ Interface," in *The Physics and Chemistry of SiO$_2$ and the Si-SiO$_2$ Interface - 2* (C. R. Helms and B. E. Deal, eds.), New York: Plenum Press, 1988.

[29] E. Rosencher, A. Straboni, S. Rigo, and G. Amsel, "An ^{18}O Study of the Thermal Oxidation of Silicon in Oxygen," *Appl. Phys. Lett.*, vol. 34, no. 4, pp. 254–256, 1979.

BIBLIOGRAPHY

[30] R. Singh, "Rapid Isothermal Processing," *J. Appl. Phys.*, vol. 63, no. 8, pp. R59–R114, 1988.

[31] R. J. Kriegler, "Neutralization of Na^+ Ions in HCL-Grown SiO_2," *Appl. Phys. Lett.*, vol. 20, no. 11, pp. 449–451, 1972.

[32] B. E. Deal and D. W. Hess, "Kinetics of the Thermal Oxidation of Silicon in O_2/H_2O and O_2/Cl_2 Mixtures," *J. Electrochem. Soc.*, vol. 125, no. 2, pp. 339–346, 1978.

[33] D. W. Hess and B. E. Deal, "Kinetics of the Thermal Oxidation of Silicon in O_2/HCl Mixtures," *J. Electrochem. Soc.*, vol. 124, no. 5, pp. 735–739, 1977.

[34] B. E. Deal, "Thermal Oxidation Kinetics of Silicon in Pyrogenic H_2O and 5% HCL/H_2O Mixtures," *J. Electrochem. Soc.*, vol. 125, no. 4, pp. 576–579, 1978.

[35] L. N. Lie, R. R. Razouk, and B. E. Deal, "High Pressure Oxidation of Silicon in Dry Oxygen," *J. Electrochem. Soc.*, vol. 129, no. 12, pp. 2828–2834, 1982.

[36] R. R. Razouk, L. N. Lie, and B. E. Deal, "Kinetics of High Pressure Oxidation of Silicon in Pyrogenic Steam," *J. Electrochem. Soc.*, vol. 128, no. 10, pp. 2214–2220, 1981.

[37] L. E. Katz and L. C. Kimerling, "Defect Formation during High Pressure, Low Temperature Steam Oxidation of Silicon," *J. Electrochem. Soc.*, vol. 125, no. 10, pp. 1680–1683, 1978.

[38] E. A. Lewis and E. A. Irene, "The Effect of Surface Orientation on Silicon Oxidation Kinetics," *J. Electrochem. Soc.*, vol. 134, no. 9, pp. 2332–2339, 1987.

[39] J. R. Ligenza, "Effect of Crystal Orientation on Oxidation Rates of Silicon in High Pressure Steam," *Journal of Physical Chemistry*, vol. 65, no. 11, pp. 2011–2014, 1961.

[40] D. A. Buchanan and S. H. Lo, "Reliability and Intergration of Ultra-Thin Gate Dielectrics for Advanced CMOS," *Microelectronic Engineering*, vol. 36, pp. 13–20, 1997.

[41] S. V. Hattangady, H. Niimi, and G. Lucovsky, "Controlled Nitrogen Incorporation at the Gate Oxide Surface," *Appl. Phys. Lett.*, vol. 66, no. 25, pp. 3495–3497, 1995.

[42] C. R. Helms, "Thermal Routes to Ultrathin Oxynitrides," in *Fundamental Aspects of Ultrathin Dielectrics on Si-based Devices* (E. Garfunkel, E. P. Gusev, and A. Y. Vul, eds.), pp. 181–190, Dordrecht, The Netherlandes: Kluwer Academic Publishers, 1998.

BIBLIOGRAPHY

[43] G. Lucovsky, A. Banerjee, B. Hinds, B. Claflin, K. Koh, and H. Yang, "Minimization of Suboxide Transition Regions at Si-SiO$_2$ Interfaces by 900 °C Rapid Thermal Annealing," *J. Vac. Sci. Technol. B*, vol. 15, no. 4, pp. 1074–1079, 1997.

[44] W. Ting, H. Hwang, J. Lee, and D. L. Kwong, "Growth Kinetics of Ultrathin SiO$_2$ Films Fabricated by Rapid Thermal Oxidation of Si Substrates in N$_2$O," *J. Appl. Phys.*, vol. 70, no. 2, pp. 1072–1074, 1991.

[45] Y. Okada, P. J. Tobin, K. G. Reid, R. I. Hegde, B. Maiti, and S. A. Ajuria, "Furnace Grown Gate Oxynitride using Nitric Oxide (NO)," *IEEE Trans. Electron Devices*, vol. 41, no. 9, pp. 1608–1613, 1994.

[46] E. P. Gusev, H. C. Lu, T. Gustafsson, E. Garfunkel, M. L. Green, and D. Brasen, "The Composition of Ultrathin Silicon Oxynitrides Thermally Grown in Nitric Oxide," *J. Appl. Phys.*, vol. 82, no. 2, pp. 896–898, 1997.

[47] K. A. Ellis and R. A. Buhrman, "Furnace Gas-Phase Chemistry of Silicon Oxynitridation in N$_2$O," *Appl. Phys. Lett.*, vol. 68, no. 12, pp. 1696–1698, 1996.

[48] E. P. Gusev, H. C. Lu, E. Garfunkel, T. Gustafsson, and M. L. Green, "Growth and Characterization of Ultrathin Nitrided Silicon Oxide Films," *IBM J. Res. Develop.*, vol. 43, no. 3, pp. 265–286, 1999.

[49] M. L. Green, T. Sorsch, L. C. Feldman, W. N. Lennard, E. P. Gusev, E. Garfunkel, H. C. Lu, and T. Gustafsson, "Ultrathin SiO$_x$N$_y$ by Rapid Thermal Heating of Silicon in N$_2$ at T = 760–1050 °C," *Appl. Phys. Lett.*, vol. 71, no. 20, pp. 2978–2980, 1997.

[50] I. J. Baumvol, F. C. Stedile, J. J. Ganem, I. Trimaille, and S. Rigo, "Thermal Nitridation of SiO$_2$ Films in Ammonia: The Role of Hydrogen," *J. Electrochem. Soc.*, vol. 143, no. 4, pp. 1426–1434, 1996.

[51] B. E. Deal and A. S. Grove, "General Relationship for the Thermal Oxidation of Silicon," *J. Appl. Phys.*, vol. 36, no. 12, pp. 3770–3778, 1965.

[52] L. Pauling, "The Nature of Silicon-Oxygen Bonds," *American Mineralogist*, vol. 65, pp. 321–323, 1980.

[53] Wikipedia, *Henry's Law*. http://en.wikipedia.org/wiki/Henry's_law.

[54] Y. J. van der Meulen, "Kinetics of Thermal Growth of Ultra-Thin Layers of SiO$_2$ on Silicon: Experiment," *J. Electrochem. Soc.*, vol. 119, no. 4, pp. 530–534, 1972.

[55] S. M. Hu, "New Oxide Growth Law and the Thermal Oxidation of Silicon," *Appl. Phys. Lett.*, vol. 42, no. 10, pp. 872–874, 1983.

[56] H. Z. Massoud, J. D. Plummer, and E. A. Irene, "Thermal Oxidation of Silicon in Dry Oxygen Growth Rate Enhancement in the Thin Regime: Experimental Results," *J. Electrochem. Soc.*, vol. 132, no. 11, pp. 2685–2693, 1985.

[57] H. Z. Massoud, J. D. Plummer, and E. A. Irene, "Thermal Oxidation of Silicon in Dry Oxygen: Accurate Determination of the Kinetic Rate Constants," *J. Electrochem. Soc.*, vol. 132, no. 7, pp. 1746–1753, 1985.

[58] H. Z. Massoud and J. D. Plummer, "Analytic Relationship for the Oxidation of Silicon in Dry Oxygen in the Thin-Film Regime," *J. Appl. Phys.*, vol. 62, no. 8, pp. 3416–3423, 1987.

[59] C. P. Ho, J. D. Plummer, S. E. Hansen, and R. W. Dutton, "VLSI Process Modeling—SUPREM III," *IEEE Trans. Electron Devices*, vol. 30, no. 11, pp. 1438–1453, 1983.

[60] D. Chin, S. Y. Oh, S. M. Hu, R. W. Dutton, and J. L. Moll, "Two-Dimensional Oxidation," *IEEE Trans. Electron Devices*, vol. 30, no. 7, pp. 744–749, 1983.

[61] M. E. Law, "Grid Adaption Near Moving Boundaries in Two Dimensions for IC Process Simulation," *IEEE Trans. Computer-Aided Design*, vol. 14, no. 10, pp. 1223–1230, 1995.

[62] D. Chin, S. Y. Oh, and R. W. Dutton, "A General Solution Method for Two-Dimensional Nonplanar Oxidation," *IEEE Trans. Electron Devices*, vol. 30, no. 9, pp. 993–998, 1983.

[63] V. Senez, S. Bozek, and B. Baccus, "3-Dimensional Simulation of Thermal Diffusion and Oxidation Processes," *IEDM Technical Digest*, pp. 705–708, 1996.

[64] H. Matsumoto and N. Fukuma, "Numerical Modeling of Nonuniform Si Thermal Oxidation," *IEEE Trans. Electron Devices*, vol. 32, no. 2, pp. 132–140, 1985.

[65] S. Cea, *Multidimensional Viscoelastic Modeling of Silicon Oxidation and Titanium Silicidation*. PhD thesis, University of Florida, Gainesville, 1996.

[66] U. Weinert and E. Rank, "A Simulation System for Diffuse Oxidation of Silicon: One-Dimensional Analysis," *Zeitschrift für Naturforschung A*, vol. 46, no. 11, pp. 955–966, 1991.

[67] E. Rank and U. Weinert, "A Simulation System for Diffuse Oxidation of Silicon: A Two-Dimensional Finite Element Approach," *IEEE Trans. Computer-Aided Design*, vol. 9, no. 5, pp. 543–550, 1990.

[68] F. J. Norton, "Permeation of Gaseous Oxygen through Vitreous Silica," *Nature*, vol. 191, p. 701, 1961.

BIBLIOGRAPHY

[69] A. J. Moulson and J. P. Roberts, "Water in Silica Glass," *Transactions of the Farady Society*, vol. 57, pp. 1208–1216, 1961.

[70] D. Gross, W. Hauger, W. Schnell, and P. Wriggers, *Technische Mechanik 4: Hydromechanik, Elemente der Höheren Mechanik, Numerische Methoden*. Berlin: Springer Verlag, 4th ed., 2002.

[71] F. Ziegler, *Technische Mechanik der Festen und Flüssigen Körper*. Wien: Springer Verlag, 2nd ed., 1992.

[72] R. P. Feynman, R. B. Leighton, and M. Sands, *Lectures on Physics, Volume II*. Reading, MA: Addison-Wesley, 4th ed., 1977.

[73] C. S. Rafferty, *Stress Effects in Silicon Oxidation - Simulation and Experiments*. PhD thesis, Stanford University, California, 1990.

[74] H. Matsumoto and M. Fukuma, "A Two-Dimensional Si Oxidation Model including Viscoelasticity," in *Proc. International Electron Device Meeting (IEDM)*, pp. 39–42, 1983.

[75] S. Zelenka, *Stress Related Problems in Process Simulation*. PhD thesis, Swiss Federal Institute of Technology, Zurich, 2000.

[76] V. Senez, D. Collard, P. Ferreira, B. Baccus, M. Brault, and J. Lebailly, "Analysis and Application of a Viscoelastic Model for Silicon Oxidation," *J. Appl. Phys.*, vol. 76, no. 6, pp. 3285–3296, 1994.

[77] J. Peng, D. Chidambarrao, and G. R. Srinivasan, "Novel: A Nonlinear Viscoelastic Model for Thermal Oxidation of Silicon," *COMPEL - The International Journal for Computation and Mathematics in Electrical and Electronic Engineering*, vol. 10, no. 4, pp. 341–353, 1991.

[78] V. Senez, D. Collard, P. Ferreira, and B. Baccus, "Two-Dimensional Simulation of Local Oxidation of Silicon: Calibrated Viscoelastic Flow Analysis," *IEEE Trans. Elect. Dev.*, vol. 43, no. 5, pp. 720–731, 1996.

[79] G. Schumicki and P. Seegebrecht, *Prozeßtechnologie*. Berlin: Springer, 1991.

[80] B. Hoppe, *Mikroelektronik 2*. Würzburg: Vogel Verlag, 1998.

[81] S. T. Dunham, "A Quantitative Model for the Coupled Diffusion of Phosphorus and Point Defects in Silicon," *J. Electrochem. Soc.*, vol. 139, no. 9, pp. 2628–2635, 1992.

[82] S. T. Dunham, A. H. Gencer, and S. Chakravarathi, "Modeling of Dopant Diffusion in Silicon," *IEICE Trans. Electron.*, vol. 82, no. 6, pp. 800–812, 1998.

[83] D. A. Antoniadis, M. Rodoni, and R. W. Dutton, "Impurity Redistribution in SiO_2-Si during Oxidation: A Numerical Solution Including Interfacial Fluxes," *J. Electrochem. Soc.*, vol. 126, no. 11, pp. 1939–1945, 1979.

[84] A. Poncet, "Finite-Element Simulation of Local Oxidation of Silicon," *IEEE Trans. Computer-Aided Design*, vol. 4, no. 1, pp. 41–53, 1985.

[85] H. R. Schwarz, *Methode der Finiten Elemente*. Stuttgart: Teubner, 3rd ed., 1991.

[86] E. B. Becker, G. F. Carey, and J. T. Oden, *Finite Elements, An Introduction, Vol. 1*. Englewood Cliffs: Prentice-Hall, 1981.

[87] W. Ritz, "Über eine neue Methode zur Lösung gewisser Variationsprobleme in der mathematischen Physik," *Journal für reine und angewandte Mathematik*, vol. 135, pp. 1–61, 1909.

[88] K. J. Bathe, *Finite Elemente Methoden*. Berlin: Springer Verlag, 2nd ed., 2002.

[89] G. Kämmel, H. Franeck, and H. G. Recke, *Einführung in die Methode der Finiten Elemente*. München: Carl Hanser Verlag, 1988.

[90] R. E. White, *An Introduction to the Finite Element Method with Applications to Nonlinear Problems*. New York: Wiley, 1985.

[91] A. Kost, *Numerische Methoden in der Berechnung elektromagnetische Felder*. Berlin: Springer Verlag, 1994.

[92] G. Strang, *Applied Mathematics and Scientific Computing*. Wellesley: Wellesley-Cambridge Press, 2007.

[93] Wikipedia, *Numerical Ordinary Differential Equations*.
http://en.wikipedia.org/wiki/Numerical_ordinary_differential_equations.

[94] J. Betten, *Finite Elemente für Ingenieure 1*. Berlin: Springer Verlag, 2003.

[95] O. C. Zienkiewicz, *The Finite Element Method, Vol. 1*. London: McGraw - Hill, 4th ed., 1989.

[96] J. M. Ortega and W. C. Rheinboldt, *Iterative Solution of Nonlinear Equations in several Variables*. San Diego: Academic Press, 1970.

[97] C. Überhuber, *Computernumerik*. Berlin: Springer Verlag, 2nd ed., 1995.

[98] R. Klima, *Three-Dimensional Device Simulation with Minimos-NT*. Dissertation, Institute for Microelectronics, Vienna University of Technology, 2002.
http://www.iue.tuwien.ac.at/phd/klima/.

BIBLIOGRAPHY

[99] T. Binder, *Rigoros Integration of Semiconductor Process and Device Simulatior.* Dissertation, Institute for Microelectronics, Vienna University of Technology, 2002. http://www.iue.tuwien.ac.at/phd/binder/.

[100] S. Wagner, "The Minimos-NT Linear Equation Solving Module," Diplomarbeit, Institute for Microelectronics, Vienna University of Technology, 2001.

[101] H. J. Dirschmid, *Mathematische Grundlagend der Elektrotechnik.* Braunschweig: Vieweg, 4th ed., 1992.

[102] S. Wagner, T. Grasser, C. Fischer, and S. Selberherr, "An Advanced Equation Assembly Module," *Engineering with Computers*, vol. 21, pp. 151–163, 2005.

[103] R. Bauer, R. Sabelka, and C. Harlander, *The Smart Analysis Programs, User's Manual for Version 2.0.* Institute for Microelectronics, Vienna University of Technology, 1999.

[104] W. Wessner, H. Ceric, C. Heitzinger, A. Hössinger, and S. Selberherr, "Anisotropic Mesh Adaption Governed by a Hessian Matrix Metric," in *Proc. 15th European Simulation Symposium (ESS)*, pp. 41–46, 2003.

[105] P. Fleischmann, *Mesh Generation for Technology CAD in Three Dimensions.* Dissertation, Institute for Microelectronics, Vienna University of Technology, 1999. http://www.iue.tuwien.ac.at/phd/fleischmann/.

[106] The GNU Triangulated Surface Library, *The GTS Library*, 2006. http://gts.sourceforge.net/.

[107] A. Hössinger, J. Cervenka, and S. Selberherr, "A Multistage Smoothing Algorithm for Coupling Cellular and Polygonal Datastructures," in *Proc. Int. Conference on the Simulation of Semiconductor Processes and Devices (SISPAD)*, pp. 259–262, 2003.

[108] S. Wagner, S. Holzer, R. Strasser, R. Plasum, T. Grasser, and S. Selberherr, *SIESTA - The Simulation Environment for Semiconductor Technology Analysis.* Institute for Microelectronics, Vienna University of Technology, 2003. http://www.iue.tuwien.ac.at/software.html.

[109] D. B. Kao, J. P. McVittie, W. D. Nix, and K. C. Saraswat, "Two-Dimensional Thermal Oxidation of Silicon - II. Modeling Stress Effects in Wet Oxides," *IEEE Trans. Electron Devices*, vol. 35, no. 1, pp. 25–37, 1988.

[110] P. Sutardja and W. G. Oldham, "Modeling of Stress Effects in Silicon Oxidation," *IEEE Trans. Electron Devices*, vol. 36, no. 11, pp. 2415–2421, 1989.

[111] H. Umimoto and S. Odanaka, "Three-Dimensional Numerical Simulation of Local Oxidation of Silicon," *IEEE Trans. Electron Devices*, vol. 38, no. 3, pp. 505–511, 1991.

[112] P. Ferreira, V. Senez, and B. Baccus, "Mechanical Stress Analysis of a LDD MOSFET Structure," *IEEE Trans. Electron Devices*, vol. 43, no. 9, pp. 1525–1532, 1996.

[113] A. S. Oates, "Electromigration Failure of Contacts and Vias in Sub-Mircon Integrated Curcuit Metallizations," *Microelectronics Reliability*, vol. 36, no. 7, pp. 925–953, 1996.

[114] D. Dalleau and K. Weide-Zaage, "Three-Dimensional Voids Simulation in Chip Metallization Structures: a Contribution to Reliability Evaluation," *Microelectronics Reliability*, vol. 41, no. 9, pp. 1625–1630, 2001.

[115] D. N. Bhate, A. Kumar, and A. F. Bower, "Diffuse Interface Model for Electromigration and Stress Voiding," *J. Appl. Phys.*, vol. 87, no. 4, pp. 1712–1721, 2000.

[116] I. A. Blech, "Electromigration in Thin Aluminium Films on Titanium Nitride," *J. Appl. Phys.*, vol. 47, no. 4, pp. 1203–1208, 1976.

[117] D. R. Fridline and A. F. Bower, "Influence of Anisotropic Surface Diffusivity on Electromigration Induced Void Migration and Evolution," *J. Appl. Phys.*, vol. 85, no. 6, pp. 3168–3174, 1999.

[118] H. Ceric, R. Sabelka, S. Holzer, W. Wessner, S. Wagner, T. Grasser, and S. Selberherr, "The Evolution of the Resistance and Current Density During Electromigration," in *Proc. Int. Conference on the Simulation of Semiconductor Processes and Devices (SISPAD)*, pp. 331–334, 2004.

[119] M. A. Meyer, M. Herrmann, E. Langer, and E. Zschech, "In Situ SEM Observation of Electromigration Phenomena in Fully Embedded Copper Interconnect Structures," *Microelectronic Engineering*, vol. 64, pp. 375–382, 2002.

[120] C. K. Hu, B. Luther, F. B. Kaufman, J. Hummel, C. Uzoh, and D. J. Pearson, "Copper Interconnection Integration and Reliability," *Thin Solid Films*, vol. 262, pp. 84–92, 1995.

[121] J. R. Lloyd and J. J. Clement, "Electromigration in Copper Conductors," *Thin Solid Films*, vol. 262, pp. 135–141, 1995.

[122] M. R. Gungor, D. Maroudas, and L. J. Gray, "Effects of Mechanical Stress on Electromigration-Driven Transgranular Void Dynamics in Passivated Metallic Thin Films," *Appl. Phys. Lett.*, vol. 73, no. 26, pp. 3848–3850, 1998.

BIBLIOGRAPHY

[123] H. Ceric, *Numerical Techniques in Modern TCAD*. Dissertation, Institute for Microelectronics, Vienna University of Technology, 2005. http://www.iue.tuwien.ac.at/phd/ceric/.

[124] R. Sabelka and S. Selberherr, "A Finite Element Simulator for Three-Dimensional Analysis of Interconnect Structures," *Microelectronics Journal*, vol. 32, no. 2, pp. 163–171, 2001.

[125] C. Harlander, R. Sabelka, R. Minixhofer, and S. Selberherr, "Three-Dimensional Transient Electro-Thermal Simulation," in *Proc. Therminic Workshop*, pp. 169–172, 1999.

[126] R. Sabelka, *Dreidimensionale Finite Elemente Simulation von Verdrahtungsstrukturen auf Integrierten Schaltungen*. Dissertation, Institute for Microelectronics, Vienna University of Technology, 2001. http://www.iue.tuwien.ac.at/phd/sabelka/.

[127] D. Ang and R. V. Ramanujan, "Hydrostatic Stress and Hydrostatic Stress Gradients in Passivated Copper Interconnects," *Materials Science and Engineering A*, vol. 423, pp. 157–165, 2006.

[128] K. Hoshino, H. Yagi, and H. Tsuchikawa, "TiN-Encapsulated Copper Interconnects for ULSI Application," in *Proc. IEEE 6th VLSI Multilevel Interconnections Conference (VMIC)*, pp. 226–232, 1989.

[129] K. Weide, X. Yu, and F. Menhorn, "Finite Element Investigations of Mechanical Stress in Metallization Structures," *Microelectronics Reliability*, vol. 36, no. 11/12, pp. 1703–1706, 1996.

[130] D. O. Thompson and D. K. Holmes, "Dislocation Contribution to the Temperature Dependence of the Internal Friction and Young's Modulus of Copper," *J. Appl. Phys.*, vol. 30, no. 4, pp. 525–541, 1959.

[131] T. Thonhauser, T. J. Scheidemantel, J. O. Sofo, J. V. Badding, and G. D. Mahan, "Thermoelectric Properties of Sb_2Te_3 under Pressure and Uniaxial Stress," *Physical Review B*, vol. 68, no. 085210, pp. 1–8, 1996.

[132] A. Witvrouw, M. Gromova, A. Mehta, S. Sedky, P. D. Moor, K. Baert, and C. van Hoof, "Poly-SiGe, a Superb Material for MEMS," *Materials Research Society Symposium Proceeding*, vol. 782, pp. A2.1.1–A2.1.12, 2004.

[133] M. F. Dorner and W. D. Nix, "Stresses and Deformation Processes in Thin Films on Substrates," *CRC Critical Reviews in Solid State and Materials Sciences*, vol. 14, no. 3, pp. 225–267, 1988.

[134] G. Russo and P. Smereka, "A Level-Set Method for the Evolution of Faceted Crystals," *SIAM J. Sci. Comp*, vol. 21, no. 6, pp. 2073–2095, 2000.

[135] P. Smereka, X. Li, G. Russo, and D. J. Srolovitz, "Simulation of Faceted Film Growth in Three Dimensions: Microstructure, Morphology and Texture," *Acta Materialia*, vol. 53, pp. 1191–1204, 2005.

[136] B. W. Shelton, A. Rajamani, A. Bhandari, E. Chason, S. K. Hong, and R. Beresford, "Competition between Tensile and Compressive Stress Mechanisms during Volmer-Weber Growth of Aluminium Nitride Films," *J. Appl. Phys.*, vol. 98, no. 043509, 2005.

[137] P. G. Shewmon, *Transformation in Metals*. McGraw-Hill, New York, 1969.

[138] T. van der Donck, J. Boost, C. Rusu, K. Baert, C. van Hoof, J. P. Celis, and A. Witvrouw, "Effect of Deposition Parameters on the Stress Gradient of CVD and PECVD poly-SiGe for MEMS Applications," in *Proc. of SPIE - Micromachining and Microfabrication Process Technology IX*, vol. 5342, (San Jose, USA), pp. 8–18, 2004.

[139] R. W. Hoffman, "Stresses in Thin Films: The Relevance of Grain Boundaries and Impurities," *Thin Solid Films*, vol. 34, pp. 185–190, 1976.

[140] K. Cholevas, N. Liosatos, A. E. Romanov, M. Zaiser, and E. C. Aifantis, "Misfit Dislocation Patterning in Thin Films," *Physica Status Solidi (B)*, vol. 209, no. 10, pp. 295–304, 1998.

[141] E. Klokholm and B. S. Berry, "Intinsic Stress in Evaporated Metal Films," *J. Electrochem. Soc.*, vol. 115, no. 8, pp. 823–826, 1968.

[142] P. Chaudhari, "Grain Growth and Stress Relief in Thin Films," *J. Vac. Sci. Techn.*, vol. 9, no. 1, pp. 520–522, 1972.

[143] B. W. Sheldon, A. Ditkowski, R. Beresford, E. Chason, and J. Rankin, "Intinsic Compressive Stress in Polycrystalline Films with Negligible Grain Boundary Diffusion," *J. Appl. Phys.*, vol. 94, no. 2, pp. 948–957, 2003.

[144] E. Chason, B. W. Sheldon, and L. B. Freund, "Origin of Compressive Residual Stress in Polycrystalline Thin Films," *Physical Review Letters*, vol. 88, no. 15, p. 156103, 2002.

[145] L. B. Freund and E. Chason, "Model for Stress Generated upon Contact of Neighboring Islands on the Surface of a Substrate," *J. Appl. Phys.*, vol. 89, no. 9, pp. 4866–4873, 2001.

[146] S. V. Bobylev and I. A. Ovidko, "Faceted Grain Boundaries in Polycrystalline Films," *Physics of the Solid State*, vol. 45, no. 10, pp. 1926–1931, 2003.

BIBLIOGRAPHY

[147] A. Molfese, A. Mehta, and A. Witvrouw, "Determination of Stress Profile and Optimization of Stress Gradient in PECVD Poly-SiGe Films," *Sensors and Actuators A*, vol. 118, no. 2, pp. 313–321, 2005.

VDM Verlagsservicegesellschaft mbH

Die VDM Verlagsservicegesellschaft sucht für wissenschaftliche Verlage abgeschlossene und herausragende

Dissertationen, Habilitationen, Diplomarbeiten, Master Theses, Magisterarbeiten usw.

für die kostenlose Publikation als Fachbuch.

Sie verfügen über eine Arbeit, die hohen inhaltlichen und formalen Ansprüchen genügt, und haben Interesse an einer honorarvergüteten Publikation?

Dann senden Sie bitte erste Informationen über sich und Ihre Arbeit per Email an *info@vdm-vsg.de*.

Sie erhalten kurzfristig unser Feedback!

VDM Verlagsservicegesellschaft mbH
Dudweiler Landstr. 99
D - 66123 Saarbrücken

Telefon +49 681 3720 174
Fax   +49 681 3720 1749

www.vdm-vsg.de

Die VDM Verlagsservicegesellschaft mbH vertritt

Printed by Books on Demand GmbH, Norderstedt / Germany